Nanopore Sequencing
An Introduction

Nanopore Sequencing

An Introduction

Daniel Branton
Harvard

David Deamer
UC Santa Cruz

World Scientific

NEW JERSEY · LONDON · SINGAPORE · BEIJING · SHANGHAI · HONG KONG · TAIPEI · CHENNAI · TOKYO

Published by

World Scientific Publishing Co. Pte. Ltd.

5 Toh Tuck Link, Singapore 596224

USA office: 27 Warren Street, Suite 401-402, Hackensack, NJ 07601

UK office: 57 Shelton Street, Covent Garden, London WC2H 9HE

Library of Congress Cataloging-in-Publication Data

Names: Branton, Daniel, author, editor. | Deamer, David W., 1939–   author, editor.
Title: Nanopore sequencing : an introduction / by (authors), Daniel Branton, David Deamer.
Description: New Jersey : World Scientific, 2018. | Includes bibliographical references and index.
Identifiers: LCCN 2018049399 | ISBN 9789813270602 (hard cover : alk. paper)
Subjects: | MESH: Nanopores | Sequence Analysis
Classification: LCC QP623 | NLM QT 36.5 | DDC 572.8/8--dc23
LC record available at https://lccn.loc.gov/2018049399

British Library Cataloguing-in-Publication Data
A catalogue record for this book is available from the British Library.

For any available supplementary material, please visit
https://www.worldscientific.com/worldscibooks/10.1142/10995#t=suppl

Typeset by Stallion Press
Email: enquiries@stallionpress.com

Foreword

Nanopore sequencing of DNA became available to interested researchers when the MinION nanopore device was delivered to early users in 2014. Over the next four years, the long reads, portability, rapid throughput, and low cost of the MinION allowed unprecedented access to DNA and RNA base sequences in conditions from remote African villages during the Ebola epidemic to the International Space Station. Furthermore, nanopore sequencing technology is rapidly being expanded to a variety of instruments both large and small, ranging from the PromethION designed to sequence several human genomes per day to the SmidgION that can be used with a cell phone.

One of the editors of this book (DB) decided to test the MinION as a way to intensify the intellectual experience of incoming students who are considering biology as a field of concentration. We often have students who say "I want to major in Biology. Why do I have to take a course in [Physics] [Math]…etc." The book directly addresses this question by illustrating the range of topics — from electricity to biochemistry, protein structure, molecular engineering, and informatics — that are foundations of nanopore sequencing technology. By the end of the lab course, students will recognize that they accomplished something new, something the professor teaching the course probably could not have done when she or he had been a student. By sequencing DNA samples and using bioinformatics to compare the results to a database, they identified which microorganisms were growing in their petri dishes.

The course was successful, but revealed an obvious need for a textbook, which was the inspiration for writing this book. We also realized from our own experience that thousands of MinIONs are being delivered to research laboratories around the world, but there was little in the way of comprehensive explanations of how and why nanopore sequencing works. This book covers the important elements a novice must grasp in order to understand the biophysical basis for nanopore sequencing. Because the foundation knowledge required to perform nanopore sequencing is so broad, it was essential to invite specialists to write individual chapters that cover each of the areas. The authors of chapters are experienced users of nanopore sequencing and include several scientists who developed the technology, so this book will satisfy the need for expert descriptions of how nanopore sequencing works.

Anyone newly undertaking a project that requires sequencing DNA or RNA will quickly realize that there are several competing technologies, each with its own advantages and limitations, and that they need to choose which is most appropriate for the work they intend to do. Although this book focuses on nanopore sequencing, it is worth comparing it with two other methods in terms of advantages and limitations. The most highly developed current technology is based on "sequencing by synthesis". In the instruments marketed by Illumina, for instance, tens of millions of short fragments of the DNA to be analyzed are attached to the surface of a flow cell. The fragments are templates for DNA synthesis, and the flow cell is then flushed in cycles with solutions containing a polymerase enzyme and one of the four nucleoside triphosphates that is a substrate for the polymerase. Each of the four substrates — ATP, TTP, GTP, and CTP — is labeled with a fluorescent dye. After the nucleotide has been added to the template strand by the enzyme, the fluorescence from the dye is imaged and recorded by a sensitive camera. The dye is then cleaved by an enzyme so that the next base can be added. Each cycle shows the fluorescence signal for the addition of one of the four possible nucleotides, and from this the base sequences of the templates can be deduced. Sequencing by synthesis is a high-throughput approach because information is collected from tens of millions of DNA fragments in parallel. Powerful computational analysis then assembles the individual base sequences into the original genome from which the DNA was obtained.

The Illumina instruments dominate the market and have provided highly accurate DNA sequences for organisms ranging from viral and bacterial genomes to the thousands of human genomes now available for research and diagnostic applications. There are several limitations that must be taken into consideration. The DNA samples must be amplified by the polymerase chain reaction, which can introduce copying errors, and the fragments analyzed are relatively short, just hundreds of bases in length, which means that assembly into complete genomes can be computationally demanding. Furthermore, portions of genomes such as the telomeres are composed of thousands of repeating segments such as the TTAGGG sequence in the human genome. Although these can be sequenced, they cannot be reassembled into the contiguous genome from short fragments, which means that the human genome is still incomplete. Finally, the Illumina instruments are expensive, ranging from $49,000 to over a million dollars. In order to reduce costs of sequencing, they are usually installed in sequencing centers that process DNA samples from multiple sources.

Another version of sequencing by synthesis was developed by Pacific Biosciences (PacBio) which they refer to as single-molecule real-time sequencing, or SMRT for short. Instead of attaching short DNA fragments to a flow cell, the SMRT methods uses a transparent plate with a microscopic array of pits that function as zero-mode waveguides (ZMW). A waveguide does not allow light from a source to be emitted in all directions, but instead collects and guides the light to a sensor that visualizes the entire array. The SMRT technology uses a mixture of all four nucleotide substrates, each labeled with a different fluorescent tag. A single polymerase enzyme is attached to the bottom of each pit along with the DNA template to be analyzed. When a nucleotide is added to the target DNA by the polymerase, the flash of light from the tag is captured and recorded, and the sequence of flashes of different colors corresponds to the sequence of bases in the target DNA. The tag falls off during addition of the nucleotide, then leaves the 20 zL volume of the ZMW and no longer contributes to the fluorescence being monitored. The SMRT method has the advantage of providing quite long reads. Average reads are 15,000 bases long, with some exceeding 100,000 bases. The base calling error of a single read is ~11%, but with 50-fold coverage the error of the consensus is much reduced, and PacBio claims accuracy of 99.999%.

Other sequencing instruments on the market include the Ion Torrent and Thermofisher SOLID, which use different technological approaches. For our purposes, these need not be described, but should be included in comparisons of accuracy, read length, initial costs, and cost per million bases of sequence, which are the primary concerns of research laboratories when deciding which instrument to buy.

The MinION, developed and marketed by Oxford Nanopore Technologies (ONT), is the focus of this book, so it will not be described in the foreword, but is worth comparing it and nanopore sequencing more generally to other devices. Nanopore sequencing has several advantages over sequencing by synthesis methods. Small instruments such as the SmidgION or MinION can be sold with costs under $1000, compared with $49,000 for the least expensive Illumina instrument. It has by far the longest reads, averaging over 100,000 bases with continuous million base reads now being reported. Importantly, it reads DNA directly rather than requiring amplification by PCR during which base modification information is lost. And nanopore sequencing can directly read the nucleobases in both DNA and RNA. It is the only approach that can readily be made as a portable device that plugs into a user's laptop or cell phone. These qualities make it ideal for field work in remote locations when, for example, rapidly obtaining genome sequences during disease outbreaks is crucial for clarifying patterns of virus evolution, monitoring the validity of diagnostic assays, and investigating and hindering transmission chains. Low cost and portability also make nanopore sequencing ideal for laboratory courses that aim to introduce sequencing to undergraduate students.

Nanopore sequencing currently has limitations as well. For instance, it does not yet achieve the accuracy of the more expensive devices, but with 30-fold multiple reads does have >99% accuracy in comparing a bacterial genome to a reference genome. The accuracy is constantly improving with small changes to DNA preparation chemistry and major improvements to base calling methods. We note that another milestone was reached in 2018 when several laboratories combined multiple reads to sequence a human genome entirely with MinIONs. Finally, although the MinION is very inexpensive, the disposable flow cells cost $500 and provide sequence information of several gigabases before they must be recycled. This means that the cost per million bases sequenced is similar to competing instruments that perform sequencing in dedicated centers.

While this book was being written, we reminded authors that it was intended for students and new users, not for professional researchers with extensive experience in sequencing. We assumed that the level of writing should be understood by students entering a major university with a background of high school courses that introduce DNA structure and function, as well as the central theme of molecular biology, that DNA makes RNA makes protein. We did not assume that students necessarily had learned anything about protein structure, and therefore recommend they should read Chapters 1 and 2 of Branden and Tooze's beautifully illustrated book "Protein Structure" which provides a very clear and succinct description of secondary and tertiary structure in the context of hydrophilicity and hydrophobicity, as emphasized in Chapters 3 and 4 of this book. Students today also use the internet as a primary source of information, and YouTube has remarkable video presentations of biomolecular structure and function as well as basic information about how DNA sequencing can be performed. The ONT website also has brief videos illustrating nanopore sequencing.

Contents

List of Contributors

Daniel Branton, Professor of Biology
The Biological Laboratories, Harvard University
16 Divinity Avenue
Cambridge, MA 02138, USA
dbranton@harvard.edu

Alicia K. Byrd, Instructor
Biochemistry and Molecular Biology
University of Arkansas for Medical Sciences
4301 W. Markham St., Slot 516
Little Rock, AR 72205, USA
akbyrd@uams.edu

James Clarke, VicePresident of Platform Technology
Oxford Nanopore Technologies, Ltd
Gosling Building, Oxford Science Park
Edmund Halley Road
Oxford, OX4 4DQ, UK
james.clarke@nanoporetech.com

David W. Deamer, Research Professor, Biomolecular Engineering
Baskin School of Engineering
University of California, Santa Cruz
Santa Cruz, CA 95060, USA
deamer@soe.ucsc.edu

Stephen J. Fleming, Computational Scientist
The Broad Institute
415 Main St.
Cambridge, MA 02142, USA
stephenfleming@fas.harvard.edu

Adrew Heron, Director of Advanced Research
Oxford Nanopore Technologies, Ltd
Gosling Building, Oxford Science Park
Edmund Halley Road
Oxford, OX4 4DQ, UK
andy.heron@nanoporetech.com

Miten Jain, PostDoctoral Scholar
Baskin School of Engineering
University of California, Santa Cruz
Santa Cruz, CA 95060, USA
miten@soe.ucsc.edu

Nick Loman, Professor of Microbial Genomics and Bioinformatics
School of Biosciences
University of Birmingham
Edgbaston, Birmingham B15 2TT, UK
n.j.loman@bham.ac.uk

Joshua Quick, Bioinformatician and Doctoral Researcher
School of Biosciences
University of Birmingham
Edgbaston, Birmingham B15 2TT, UK
j.quick@bham.ac.uk

Kevin D. Raney, Professor and Chair
Biochemistry and Molecular Biology
University of Arkansas for Medical Sciences
4301 W. Markham St., Slot 516
Little Rock, AR 72205, USA
raneykevind@uams.edu

Chapter 1

The Development of Nanopore Sequencing

Daniel Branton and David Deamer

Over the last 60 years, many researchers have focused their efforts on developing methods to determine the sequence of nucleotides in genomic DNA. One of the first major breakthroughs that altered the future course of DNA sequencing technology came in 1975–1977 with Frederick Sanger's chain-termination or "dideoxy" method (Figure 1.1). This technique employs a DNA polymerase to synthesize new DNA *in vitro* using strands of the unknown sequence as templates — basically a "sequencing by synthesis" method. Synthesis is conducted in the presence of the 4 deoxyribonucleotides (dNTPs) of A, C, G, or T, plus a lesser concentration of the 4 dideoxynucleotides (ddNTP) of A, C, G, or T (Figure 1.1(a)). Because the dideoxynucleotides lack a hydroxyl group that is required for extension of the DNA chains, whenever a polymerase randomly incorporates a ddNTP into the newly synthesized strand instead of the analogous dNTP, further synthesis is terminated. In updated incarnations of Sanger's method, each of the four dideoxynucleotides is covalently bonded to a differently colored fluorescent dye. Thus, all of the newly synthesized strands in the reaction mix which end in a given nucleobase fluoresce with the same color (Figure 1.1(b)). Separating these newly synthesized fluorescent strands using gel or capillary electrophoresis allows one to see the color of each length fragment and thus identify

a. The unknown sequences (grey letters) are ligated to a short known oligonucleotide (black letters) and then incubated with primers that base-pair with the known oligonucleotide as well as dNTPs (black A, T, C and G) plus some ddNTPs. Each of the ddNTPs is labeled with a different fluorescent die colored A, T, C, or G.

b. In the presence of active DNA polymerase, the primers are elongated by addition of the appropriate complementary nucleotides. When by chance a fluorescently labelled ddNTP is added to one of the growing strands, further polymerization terminates. Each of the four dideoxy nucleotides is attached to a differently colored fluorescent dye. Thus, all of the strands in the reaction mix which end in a given base fluoresce with the same color.

c. The different length DNA strands are electrophoretically separated and appear as differently colored bands. The short length strands move through the gel more rapidly than do the longer length strands. The separatory conditions   are selected such that strands differing by only one nucleobase can be readily distinguished.

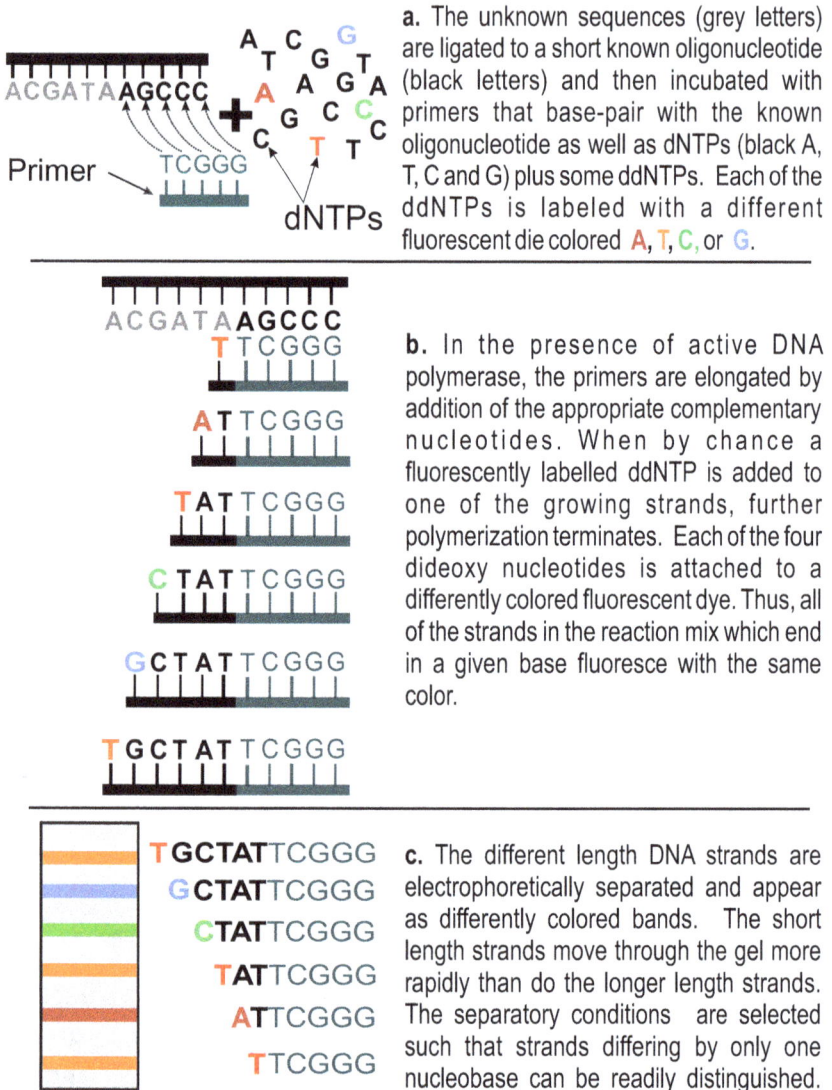

Figure 1.1.   Sanger sequencing. The three sequential steps in Sanger Sequencing, a, b, and c.

its terminal nucleotide (Figure 1.1(c)). Since the DNA fragment lengths differ from each other by one nucleotide, one can read off the sequence by simply identifying the color of each successive fragment. The electrophoresis process used with Sanger sequencing can usually read no more than 500–800 nucleotide-length fragments. To derive the sequence of a longer stretch of DNA, researchers sequence overlapping DNA fragments separately, and then use computational methods to assemble the overlaps into one long contiguous sequence read resembling the continuous length of DNA in a chromosome.

Automated versions of Sanger's dideoxy method significantly increased sequencing speed, and the first-generation machines facilitated sequencing genomes of increasingly complex species during the 1980s. Automated machines also made it realistic to plan sequencing the entire ~3,300,000,000 base pairs of the human genome 23 pairs of chromosomes. Adopted in 1990 as an international collaborative research program, the Human Genome Project produced the first finished sequence of a *Homo sapiens'* genome by 2003. One of the more surprising revelations was that the human genome contained fewer than 21,000 genes. This was far less than the ~100,000 genes estimated just a decade earlier because it turns out that over 95% of the DNA in humans does not specify a protein. We still do not know the function of this DNA, or to what extent it governs human traits!

Francis Collins, who directed the US National Human Genome Research Institute (NHGRI) as the Human Genome Project was completed, considered the sequence to be "…a transformative textbook of medicine, with insights that will give health care providers immense new powers to treat, prevent and cure disease." Collins and his colleagues at NHGRI considered that genomics would be a central and cohesive discipline underlying biomedical and agricultural research as well as evolutionary studies.[1] But establishing a robust path from genomic information to improved health remains a major challenge that requires far more knowledge about the function and interactions of animal and plant DNA than was available in 2003, or is available even today.

Progress toward the clinical use of genomic information will require comparing the phenotypes and genomes of many humans, looking for differences in DNA sequences that might be correlated with diseases. Since the statistical reliability of such correlations requires many replicate observations, it was clear in 2003 that it would be necessary to sequence many more human genomes than could affordably be completed using Sanger's dideoxy chain termination method. Even when automated, the method remained a slow and costly approach for determining the sequence of the large and complex genomes of humans and other mammals. For instance, it had cost about $3 billion and taken over 10 years to map and sequence the first human genome. Even in 2007, with a reference human genome to facilitate assembling the genome of a person whose genome had not previously been sequenced, NHGRI was still estimating a ~$10 million/genome cost.

In an effort to bring down these costs, NHGRI initiated a grant program called the $1,000 Genome that greatly encouraged both industrial enterprises and academic groups to develop next-generation methods that promised sharp reductions of the time and cost of sequencing. Although most of the commercial methods continued to use polymerase-dependent synthesis or other nucleotide-recognizing enzymatic methods, the combination of such natural biological discriminators with massively parallel procedures produced an unusually rapid pace of genome technology development. Thus, between 2007 and the end of 2015 the cost of sequencing one human genome plunged from $10,000,000 to just over $1,000.

But two important difficulties with sequencing by synthesis methods continue to be limiting. The first is that massively parallel procedures consume large amounts of genomic DNA that usually must be generated by amplifying smaller amounts of DNA from humans or other organisms. Unfortunately, all of the available amplification methodologies introduce a small number of errors and also cause statistically significant bias relative to the unamplified control. Thus, some portions of the genome may be replicated many times while other portions are poorly amplified or even left unreplicated. Such statistically significant bias may render the finished sequence unusable for certain applications. The second difficulty is that these sequencing methods produce relatively short reads that are only a few hundred nucleotides long. When a high-precision reference genome sequence exists (e.g. humans, mice) the aim of sequencing is usually to identify genetic variations, single base mutations, or insertions and deletions, that are called indels for short. These can be found by taking advantage of the overall sequence similarity among members of one species by aligning the newly determined short sequence reads with existing reference sequences.

But for *de novo* sequencing, assembling the short fragments can be a daunting task that consumes many hours of computer time. The short reads must be assembled using areas in which they overlap with each other to put together a sequence that represents the original full-length chromosome or region of interest. In either case, there are likely to be areas of the genome, particularly those containing repetitive sequences, in which it is often impossible to determine the correct sequence. The result is that important deletions, insertions, or single-nucleotide polymorphisms (SNPs, pronounced 'snips') may be overlooked.

Starting in 1996, a radically different approach was initiated in academic laboratories with the discovery that long, single strands of nucleic acids — either RNA or single-stranded DNA (ssDNA) — can be captured and electrophoretically driven through a nanopore, a nanoscale channel in a lipid bilayer separating two chambers filled with buffered potassium chloride (KCl) solutions.[2] This discovery led to three critical findings that made it possible to sequence an organism's genome using a nanopore:

(1) A simple routine to drive a genome's deoxynucleotides through a nanopore in precisely the same sequence as they occur in the organism's chromosomes;

(2) The discovery of protein nanopores in which the presence of each of the four naturally occurring deoxynucleotides modifies the pore's ionic conductivity in a different and recognizable manner;

(3) The development of methods to drive single strands of chromosomal DNA through a nanopore in single steps, each the length of a deoxynucleotide monomer.

Two significant features of nanopore sequencing are that amplification of the DNA is not usually required, and the nucleobase sequence of much longer strands can be determined. A typical setup used in the first studies that led to these findings is presented in Figures 1.2(a) and 1.2(b), which show how a DNA strand can be drawn through a nanopore by an applied voltage. The only ion-containing fluid path is through the nanopore in a lipid bilayer that is formed between two KCl-filled chambers, called *cis* and *trans*. α-hemolysin, a readily available mushroom-shaped

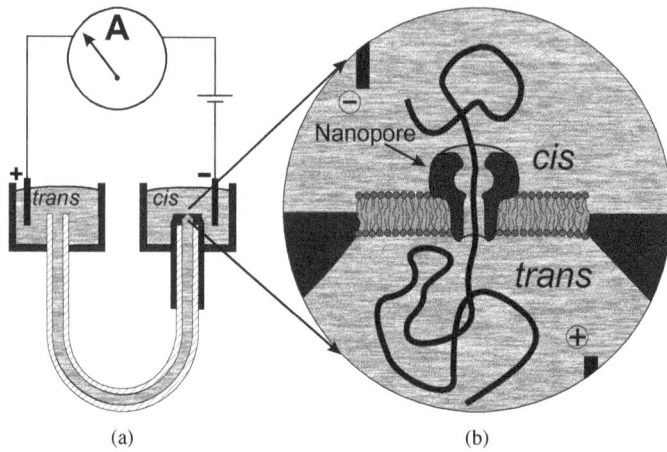

(a)   (b)

Figure 1.2. Typical setup for a nanopore experiment. (a) Two interconnected chambers called *cis* and *trans* are filled with ionic solutions of KCl. To facilitate access and provide a clear view into the *cis* chamber, a U-shaped solution-filled tube is used to connect the two KCl-filled chambers. (b) A high-magnification sketch of the end of the U-tube in the *cis* chamber that terminates with a lipid bilayer containing a nanopore. The ionic solution in the nanopore's channel provides the only aqueous connection through which ions or molecules can translocate from the solution on the *cis* side of the bilayer to the solution on the *trans* side of the bilayer. A voltage bias between the *cis* solution and the *trans solution* drives single-stranded polynucleotides placed in the *cis* chamber to translocate through the nanopore to the *trans* chamber. A sensitive ammeter, shown in (a), measures the accompanying ionic current flowing *cis* to *trans*.

protein produced by the gram-positive bacterium *Staphylococcus aureus*, was used in many of the initial experiments. As explained in Chapter 4, α-hemolysin self-assembles to form a ~1.5 nm diameter aqueous channel through the lipid bilayer. This ~1.5 nm diameter is much greater than the diameter of hydrated K^+ and Cl^- ions, which pass through easily at a rate of 700 million ions per second, but the ~1.2 nm diameter of a single polynucleotide strand of DNA can barely fit through. Double-stranded DNA (dsDNA) has twice that diameter and cannot traverse through the pore.

In a typical experiment, a voltage bias of ~120–180 mV is applied across the membrane by electrodes placed in each KCl solution. No ionic current can be measured because the bilayer acts as a barrier.

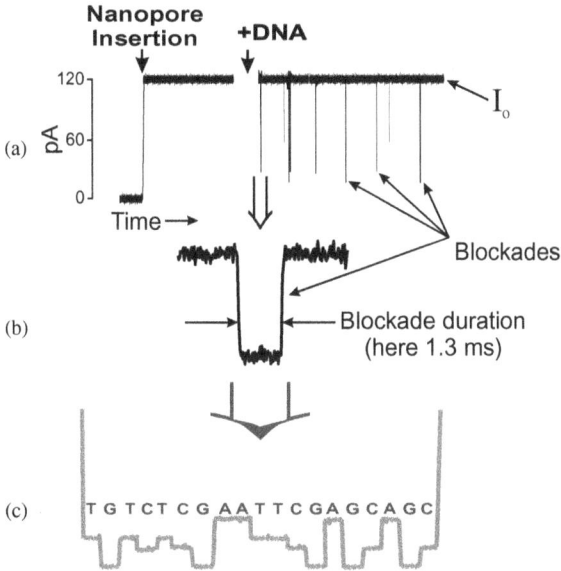

Figure 1.3. Ionic current through a nanopore setup shown in Figure 1.2. (a) A bilayer is impermeable to ions. Before a nanopore self-inserts itself through the bilayer, no ions flow *cis* to *trans* even after a voltage bias is applied *cis* to *trans*. After nanopore insertion, ions can flow and the baseline open pore current, I_o, is typically about 120 pA with a 120 mV *cis* to *trans* bias. Addition of polynucleotides to the *cis* chamber cause brief current diminutions called 'blockades'. Most of the blockades are interpreted as polynucleotide strands being electrophoresed through the nanopore. (b) One blockade. Time axis expanded x500. (c) Hypothetical view of one blockade. With better control of the polynucleotide movement through an enhanced nanopore, Deamer and colleagues hypothesized that what appeared as a single blockade would, with better resolution, prove to be a sequence of current levels that could be decoded to reveal the sequence of nucleotides translocating through the nanopore.[2]

Only after an α-hemolysin channel spontaneously inserts into the lipid bilayer is there a sudden increase in ionic current, referred to as the open channel current. Then, when single-stranded DNA or RNA is added to the *cis* side of the chamber filled with KCl, the ionic current through the channel undergoes transient reductions referred to as blockades (Figures 1.3(a) and 1.3(b)). Furthermore, small modulations of the ionic current can sometimes be observed within the blockades.

These results are interpreted as showing that after addition of nucleic acids to the *cis* chamber, when a polynucleotide diffuses very near the voltage-biased nanopore, one end of the strand enters the electric field within and surrounding that nanopore's channel. Once in this polarized field, the polynucleotide — whose sugar-phosphate backbone gives nucleic acids a negative charge — is electrophoresed through the nanopore toward the positively charged electrode contacting the KCl solution on the *trans* side of the membrane. The K^+ and Cl^- ions in solution are of course also drawn through the channel — the K^+ cations to the negatively charged electrode and the Cl^- anions to the positively charged electrode. As will be explained in Chapter 2, the movement of these ions produces an ionic current through the α-hemolysin nanopore measured in units of amperes (A). The 120 picoamperes (pA) shown in Figure 1.3 is equivalent to 700 million ions per second, a current easily measured by amplifiers that can resolve currents less that a picoampere in magnitude. As noted earlier, the ~1.5 nm diameter channel of α-hemolysin is only slightly larger than the ~1.2 nm diameter of a polynucleotide strand. The presence of a traversing polynucleotide nearly fills the nanopore's aperture and decreases the number of K^+ and Cl^- ions that can simultaneously traverse the aperture by 80–90%. This explains the diminished current flow — the blockade — during the time it takes for a polymer molecule to be fully translocated through the nanopore (Figure 1.3(b)). Readers interested in the theory underlying polymer motion and translocation should consult Muthukumar[3] for a cogent analysis of how nucleic acid strands can be driven through nanoscopic pores by an applied voltage.

The interpretation that the blockades in Figure 1.3 are caused by polynucleotides translocating through the narrow nanopore surprised many scientists who offered alternative interpretations of the data. Because a single strand of DNA or RNA is extremely flexible and usually tangles up into a disordered roughly spherical mass when in solution, it seemed improbable that the electric field near a voltage-biased nanoscale pore could successfully draw one or the other end of such a tangled mass into the nanopore's aperture. An alternative explanation for a blockade was that each time a tangled polynucleotide diffused into the electric field near a voltage biased nanopore, the tangled mass collided with or briefly seated itself on or in the nanopore's entrance aperture without actually traversing through the nanopore, thus reducing or obstructing the number of ions that could enter and flow through the nanopore's channel. Indeed, careful consideration of the current values and durations of all the current fluctuations in each experiment indicated that some of the collisions do occur without the polynucleotide entering the nanopore's aperture. But such collisions were easily distinguished from true translocations and could be identified by their brevity (<110 μs) and their smaller effect on the ionic current (<70% of the open pore current compared to 90% blockage by a nucleic acid strand traversing the channel).[2]

Unambiguous proof that all of the blockades with durations greater than 110 μs and with current reductions greater than 70% of the open pore current were due to polynucleotides being fully translocated through the nanopore to the *trans* side of the membrane was demonstrated by counting the actual number of polynucleotide molecules present in the *trans* solution. The number of molecules always corresponded to the total number of blockades *after subtracting* the easily identified collisions. This made it clear that the large ionic current reductions during a blockade were due to the presence of polynucleotides translocating through the nanopore.

Knowing that the nanopore diameter is barely greater than the diameter of a ssDNA polymer, David Deamer, who had initiated the research on nucleic acids in nanopores, surmised that the polymer's nucleotides would be constrained by the nanopore's small diameter aperture to move through this aperture in strictly single file order. Deamer and his colleagues proposed that with further improvements to the operation of the system, the small variations of ionic current within each blockade would actually reflect the sequence of nucleotides traversing through the nanopore.[2] If so, determining the sequence of nucleobases in a strand of DNA could be accomplished simply by driving polynucleotides through a nanopore and recording the modulations in current levels. The varying current levels could then be decoded into the polynucleotide's sequence, as shown in Figure 1.3(c). Unlike Sanger's dideoxy method or any of the next-generation sequencing methods that relied on biochemical and enzymatic recognition mechanisms, this suggestion for 'nanopore sequencing' proposed to use the direct interaction of nucleobases with the electrical field in a nanopore to achieve inexpensive, rapid sequencing. In effect, Deamer's 'nanopore sequencing' proposed to directly transduce the identity of DNA's nucleobases into recognizable electrical signals.

Initially, most investigators interested in DNA sequencing found this radically different nanopore sequencing proposal to be farfetched. But as further research continued to demonstrate the capacity of a nanopore's current blockades to characterize aspects of the length, composition, motion, and other features in nucleic acid polymers,[3–5] it became clear by 2010 that just two major obstacles to achieving nanopore sequencing remained to be overcome: (a) the rate of nucleotide traversal through the nanopore had to be controlled so that the small ionic current variations within each blockade could be measured; and (b) these varying current levels had to be directly related to the sequence of bases in the strand. Overcoming these two obstacles became the major focus of research activity in several laboratories.

Research to overcome the first obstacle — controlling the rate of DNA translocation through nanopores — was essential because even the smallest voltage bias that can be used to drive ssDNA through a nanopore translocates the polynucleotide so rapidly that the ionic current associated with each successive nucleobase cannot be resolved. Ideally, the DNA strand should step through the nanopore with a rachet-like motion so that as each successive nucleobase moves into the nanopore's sensing region, it remains there long enough to permit accurate measurement of the ionic current.

A logical approach to controlling ssDNA translocation through a nanopore is to attach the strand to some sort of braking device that can control movement of the DNA through the nanopore. As will be explained in Chapter 6, enzymes such as helicases and polymerases, which are often called motor proteins, use the free-energy released by hydrolysis of high-energy phosphate bonds to fuel their directional movement along a strand of DNA. Some helicases "walk" from the 3' end to the 5' end of a DNA strand while others walk in the 5' to 3' direction. But even as they walk along a ssDNA, they remain firmly bound to the strand. Furthermore, these motor proteins have too large a diameter to translocate through the nanopore. Thus, when a single motor enzyme molecule is bound to a ssDNA, it will prevent further translocation of the DNA through the nanopore once the bound motor moves into contact with the nanopore's aperture through which it cannot fit. Thereafter, the DNA can be pulled through the nanopore by the applied voltage bias only as rapidly as the motor protein steps along the DNA in a direction away from the nanopore. Depending on which of several proteins is selected, motor proteins walk along DNA strands at rates that range from 10–1,000 bases/s, while polynucleotides without a bound motor protein usually travel through a nanopores at a rate of >1,000,000 bases/s, or ~1–2 bases/μs.

Motor proteins can move along ssDNA one nucleotide at a time. This is important, because it means that the DNA is ratcheted through the pore in steps whose length corresponds precisely to the distance between each successive nucleobase in a polynucleotide strand. Both the stepping rate and the stepping distance are suitable for controlling DNA translocation for nanopore sequencing. After testing several different motor proteins, Mark Akeson and his students found that a polymerase from the *Bacillus subtilis* phi29 bacteriophage was a promising candidate.[6] When tested, phi29 polymerase provided the first clear demonstration that nanopores can read DNA sequences with single nucleotide resolution.[7]

The need for a method to overcome the second obstacle — ionic current measurements that reflect the presence of only one traversing nucleotide at a time — became evident when early work with different oligomers showed not one, but multiple nucleobases, produced the current level changes observed as a ssDNA molecule translocated through the α-hemolysin pore. That a large number of nucleobases would simultaneously affect the current through the nanopore is not surprising given that the distance from one nucleotide to the next in a single strand of DNA is <0.5 nm whereas α-hemolysin sensing region — the entire α-hemolysin stem domain – is ~5 nm long (Fig 1.4).

An obvious solution to this problem was to discover a protein that assembles into a nanopore with a more favorable, shorter sensing zone. Although no protein has yet been found with a sensing zone that reflects the presence of only 1 nucleobase in the translocating DNA strand, several

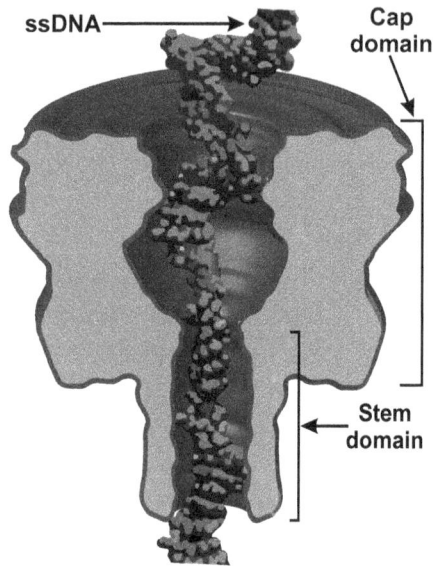

Figure 1.4.   Single-stranded DNA in α–hemolysin. All of the nucleotides translocating through α–hemolysin's stem region are sensed simultaneously because nucleobases anywhere in this stem region significantly increase the nanopore's ionic current resistance.

proteins have been shown to have a zone that reflect the presence of no more than 4–5 nucleobases, significantly fewer than with α-hemolysin. Fortunately, dynamic programming-based algorithms and other bioinformatics approaches can cope with a smaller number of nucleobases simultaneously affecting the current through the nanopore.

Inspired by nanopore results from several academic laboratories, Oxford Nanopore Technologies (ONT) developed commercially viable nanopore sequencing instruments. A MinION sequencer was the first of these to become available to the sequencing community in 2014. The MinION and subsequent instruments from ONT are based on many of the ideas and licensed patents emerging from academic laboratories as well as a large number of innovations and patents developed by the company that significantly improve the ease of use, efficiency, performance, and reproducibility for most users (Figure 1.5). For example, instead of using polymerases, DNA

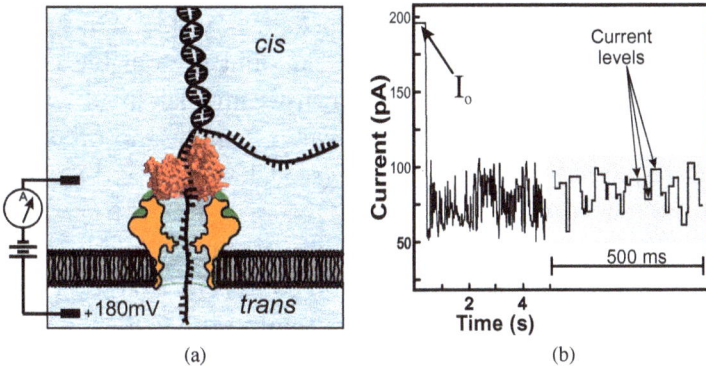

Figure 1.5. Nanopore sequencing. (a) Diagram of the functional components of each nanopore in a MinION. An ionic solution containing kilobase-fragments of genomic DNA is place in the *cis* compartment ionic solution which is separated from the *trans* compartment solution by a synthetic membrane with an embedded CsgG nanopore. An inactivated helicase molecule is bound to one of the strands in each dsDNA fragment. As this strand is drawn into the nanopore by the voltage bias across the membrane, the helicase is activated and unzips the dsDNA as it steps along the strand to which it is bound. The helicase eventually reaches the distal end of the dsDNA and falls off as the DNA strand on which it had been bound is drawn through the nanopore into the *trans* compartment. (b) A sample record of the ionic current amplitude trace observed before (I_0) and during the first ~5 s. after the DNA has been captured and travels *cis* to *trans* at ~80 bases/s through one of the nanopores in a MinION. A 500 ms sample of this output is shown with the time axis expanded by 10-fold in the grey box. The expanded time axis makes it clear that what appear to be meaningless noise-like squiggles at low time-resolution are in fact a series of ionic current levels whose nucleobase specific amplitudes can be decoded to reveal the sequence of nucleotides in the translocating DNA strand.

translocation through the nanopores is controlled by a helicase; instead of α–hemolysin, the MinION nanopore is a mutant of a bacterial amyloid secretion channel, CsgG; instead of working with one nanopore connected to separate expensive electronic components as do most academic laboratories, the 500 nanopores that work in parallel in a MinION are controlled and operated by a standard laptop computer; and instead of using PCR to amplify genomic DNA and then pre-pare short fragments for sequencing, one of several different prepar-atory kits can be used to prepare DNA without amplification in a manner that maximizes the sequence data yield. Which method and kit to be use will depend on the amount of available DNA, the sequencing accuracy required for downstream analysis, the length of each sequence read required for *de novo* assembly, and other considerations required by the investigator (see Chapters 7 and 8).

In the few years since the MinION was released to researchers, it has demonstrated several remarkable attributes. Among the most striking are its small size (~100 g) and control by a laptop computer which makes it highly portable and useful in the field. Because data is available in near real time, with the first sequenced DNA fragments ready for analysis very soon after the start of a sequencing run, a pathogenic virus or bacterium can in many cases be identified on site in under 15 min. In 2015, MinIONs were carried to Africa in a suitcase and used to follow the spread and evolution of the Ebola virus during the epidemic in West Africa. The sequencing was performed on site in the affected villages, something that could only be done with a device as small and portable as the MinION and its controlling laptop computer. And given its light weight, it was possible to ship two MinIONs to the International Space Station in July 2016 to perform the first sequencing of DNA in space. In the future, to protect astronauts from potential pathogenic bacteria and molds, MinIONs will be used to monitor the microbial populations that are present in the International Space Station.

Nucleobases in a strand of RNA that is driven through a nanopore are sensed and transduced into identifiable electrical signals as readily as are the nucleobases in a translocating strand of DNA. This makes it possible for a nanopore instrument to directly sequence native RNA molecules, including all the messenger RNA gene readouts (the transcriptome) present in a cell. RNA cannot be sequenced directly by other instruments, all of which first require transcribing the RNA into complementary DNA (cDNA) and then sequencing the cDNA. But transcribing the RNA into cDNA inevitably introduces multiple biases, incomplete transcript sequences, and other artefacts that interfere with proper characterization and quantification of transcripts. Because RNA transcript isoform expression and usage is a prominent source of variation between healthy and diseased tissues in a number of medical conditions, RNA sequencing using nanopore technology has, for example, led to a better understanding of how complexity in the breast cancer susceptibility gene (BRCA1) transcript structure causes cancer development.

The megabase-length polynucleotides that nanopores can identify in one continuous read of a DNA strand have made it possible to sequence portions of genomes containing multiple repetitive sections that cannot be sequenced by most other instruments in which continuous reads are limited to only a few hundred bases. But despite the MinION instrument's unprecedented capabilities discussed here, this device is only the first pioneering step into nanopore sequencing. It is not intended to compete with other much more expensive devices that are designed to sequence larger genomes such as those in many mammals and plants. To meet these and other sequencing demands, ONT is developing a range of nanopore sequencing instruments, some larger and massively parallel that will sequence mammalian and plant genomes as well as some compact devices designed to be used with cell phones for rapid identification of genomes anywhere and anytime.

References

1. Collins, F.S., Green, E.D., Guttmacher, A.E. & Guyer, M.S. A vision for the future of genomics research. *Nature* **422**, 835–847 (2003).
2. Kasianowicz, J.J., Brandin, E., Branton, D. & Deamer, D.W. Characterization of individual polynucleotide molecules using a membrane channel. *Proc. Natl. Acad. Sci.* **93**, 13770–13773 (1996).
3. Muthukumar, M. *Polymer Translocation*. CRC Press, Boca Raton, FL. 2011.
4. Akeson, M., Branton, D., Kasianowicz, J.J., Brandin, E. & Deamer, D.W. Microsecond time-scale discrimination among polycytidylic acid, polyadenylic acid, and polyuridylic acid as homopolymers or as segments within single RNA molecules. *Biophys. J.* **77**, 3227–3233 (1999).
5. Deamer, D. & Branton, D. Characterization of nucleic acids by nanopore analysis. *Acc. Chem. Res.* **35**, 817–825 (2002).
6. Stoddart, D., Heron, A.J., Mikhailova, E., Maglia, G. & Bayley, H. Single-nucleotide discrimination in immobilized DNA oligonucleotides

with a biological nanopore. *Proc. Natl. Acad. Sci.* **106**, 7702–7707 (2009).

7. Cherf, G.M. et al. Automated forward and reverse ratcheting of DNA in a nanopore at 5-Å precision. *Nat. Biotechnol.* **30**, 344–348 (2012).

8. Manrao, E.A. et al. Reading DNA at single-nucleotide resolution with a mutant MspA nanopore and phi29 DNA polymerase. *Nat. Biotechnol.* **30**, 349–353 (2012).

Chapter 2

Ionic Currents

Stephen Fleming

Nanopore sequencing involves an ion conducting channel or nanopore in an otherwise non-conductive membrane that separates two reservoirs of electrolyte solution. When coupled with high-resolution electronic amplification, the result is an extremely sensitive instrument for making precision measurements of ionic current through the nanopore. To understand nanopore sequencing, it is therefore important to know about the behavior of ions in solution, their response to applied electrical potentials, and how these potentials are applied across a nanopore during an experiment. It is also worthwhile to discuss the principles governing how small current signals are measured, as well as the noise sources that are likely to be encountered.

Foundational Electrostatics

Any physical object, be it a block of carbon, a copper wire, or an ionic solution of salt dissolved in water, can be characterized by its resistance to the free flow of electrical charges through itself. This resistance, R, is measured in Ohms (Ω). The resistance of an object depends on its size, shape, and composition. Intuitively, by analogy to water flowing through a pipe, objects with larger cross-sectional areas allow current to flow more easily, while objects with extended lengths impede current flow more. Resistance can be written as

$$R = \rho \frac{L}{A} \qquad (2.1)$$

where A is the object's cross-sectional area and L is its length. The resistivity, ρ, is a size-independent property, specific to a given material, and is measured in Ohm-meters ($\Omega \cdot m$). The reciprocal of the resistivity, called the conductivity, $\sigma = 1/\rho$, is another way to denote the same physical property. Conductivity is more commonly reported for electrolyte solutions. The units of conductivity are sometimes written as Siemens per meter (S/m), where Siemens are inverse Ohms. For different materials, the electrical conductivity can span well over twenty orders of magnitude. In Table 2.1, the conductivity of an electrolyte solution, 1 M potassium chloride, commonly used in nanopore experiments is compared to conductivities of everyday metals and insulators.

The flow of electrical currents is governed by Ohm's law. When an electrical potential difference, that is a voltage bias, is applied across a resistor by contacting it with electrodes, the current through the resistor is governed by Ohm's law, which is given as

$$I = V/R \qquad (2.2)$$

where I is the current, measured in Amperes (A), V is the potential difference, measured in Volts (V), and R is the resistance. Current is

Table 2.1.   **Conductivities of 1M KCl and other common materials.**[1]

Material	Conductivity (S/m) at 25°C
Silver	6.3×10^7
Carbon (amorphous)	7.3×10^4
1 M potassium chloride [KCl]	10.8
Deionized water (18 MΩ-cm)	5.5×10^{-6}
Teflon (based on DuPont datasheet)	$<10^{-16}$

a measure of the flow of charge, measured in Coulombs (C), per unit time. The electrostatic potential can be thought of as a scalar field defined to have a value in Volts at all points in space. The local change in the electrostatic potential at any given point in space is called the electric field, $\vec{E}$. The little arrow above the E indicates that the electric field has both magnitude and directionality, i.e. $\vec{E}$ is a vector field. A charged particle, such as a Cl⁻ ion, has a different potential energy when it is located at different points in space. The electric field exerts a force on the Cl⁻ ion that will drive the Cl⁻ ion to the point where it has the minimum potential energy. The force exerted by the electric field can be written as $\vec{F} = q\vec{E}$, where q represents the magnitude of the charge on which the electric field is exerting force. During nanopore sequencing, the voltage applied by the electrodes sets up an electric field that exerts a force on all of the charged ions as well as the negatively charged phosphate groups on DNA or RNA molecules, driving them through the nanopore.

Ions in Solution versus Electrons in Metals

Typically a discussion of resistors and Ohm's law brings to mind electrical conductivity in solids. In a metal, the metal atoms' valence electrons become delocalized and are free to move around. These free electrons are the charge carriers in metals. In solution, where there are no free electrons, electrical current is carried instead by ions. Ions are individual, charged atoms or molecules. Bulk matter is, for the most part, neutral due to the attraction of positive and negative charges, such as the positive potassium ion [K⁺] and the negative chloride ion [Cl⁻] in solid KCl. In solid form, K⁺ and Cl⁻ are bound tightly together and therefore unable to conduct electrical charges, but in water and some other solvents, K⁺ and Cl⁻ can be separated from one another and move about independently. In water, this phenomenon is called "hydration," and it involves the ability of water molecules to screen the individual charges, as shown in Figure 2.1, by orienting their slightly negative oxygen atoms or their slightly positive hydrogen atoms near the ion, thereby shielding the ion from its partner.

Solid salt   Hydrated ions

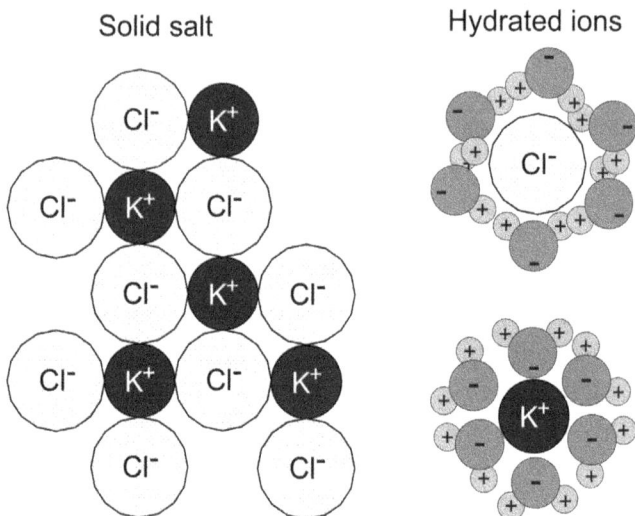

Figure 2.1.   A schematic depiction of the hydration of K^+ and Cl^- ions by water molecules, where oxygen atoms are solid light gray and hydrogen atoms are darker solid gray. To give a sense of scale, the diameter of the K^+ ion is 2.6Å and that of the Cl^- ion is 3.6Å, while a water molecule is approximately 2Å from hydrogen to hydrogen.[2]

While ions such as K^+ and Cl^- will completely dissociate in water, they will not do so in hydrophobic environments such as the hydrocarbon chains that comprise lipid tails. Unlike water, hydrocarbons do not have strong dipoles that can align around the ions to screen their mutual attraction. To get an idea of the concentration of ions in an aqueous 1 M KCl solution, consider that 1 mole of KCl (74.5 g) per liter is approximately $2 \times 6 \times 10^{23}$ ions in 0.001 m^3, which is equal to about 1 ion/nm^3. There are 55 water molecules for each potassium and chloride ion, with six or so associated with the hydrated ion.

Hydrated ions with their associated water molecules are free to move about in water. Their motions are governed by random thermal motion, called Brownian motion, which gives rise to diffusion. This omnipresent chaotic motion averages out to zero net flow of current, but if an external electric field is applied, the ions will tend to drift in certain directions, leading to a net current flow. Due to constant collisions and drag from surrounding fluid, ions do not accelerate forever.

They accelerate to a velocity where the drag friction and electrophoretic force are balanced. This velocity is described by a parameter called the mobility u, with units of velocity per electric field. The current carried by a single ion of charge q in an electric field $\vec{E}$ can be written as

$$\vec{I}_{ion} = qu\vec{E} \qquad (2.3)$$

Since this is a vector equation, positive and negative ions, with their opposite charges, will travel in opposite directions. The current density, $\vec{J}$, which can be thought of as the charge passing through a unit area per unit time, can be written as

$$\vec{J} = nq\vec{v}_d \qquad (2.4)$$

where n is the number of ions per unit volume, and $\vec{v}_d$ is the ion drift velocity. In this case, from our definition of the mobility, $\vec{v}_d = u\vec{E}$, and so

$$\vec{J} = nqu\vec{E}. \qquad (2.5)$$

In this form, the equation is a microscopic version of Ohm's law, $\vec{J} = \sigma \vec{E}$, where the conductivity is given by $\sigma = nqu$. The mobility of the potassium ion is 7.6 μm/s in a 1V/cm field, nearly the same as the chloride ion (7.9 μm/s in a 1 V/cm field). Because the potassium ion mobility is higher than that of the sodium ion (5.2 μm/s in a 1 V/cm field) or the lithium ion (4.2 μm/s in a 1 V/cm field),[2,3] the conductivity of potassium chloride is greater than either sodium chloride or lithium chloride at equal concentrations. Thus, to maximize the amount of current flowing through a nanopore, a solution of KCl, rather than NaCl or LiCl, is preferred.

DNA is also a charged molecule in solution. As the name deoxyribonucleic acid would suggest, the phosphate backbone of DNA is negatively charged in solution near neutral pH. Due to its polymeric nature, it is sometimes referred to as a polyelectrolyte. Since the DNA backbone carries one negative charge per nucleotide, it will experience a force in an electric field, just as ions do. When a potential difference is applied between two electrodes in a nanopore experiment, an electric field is set up which pulls DNA through the nanopore from one side of the membrane to the other.

Electrodes: Sources of Ionic Current

In a nanopore experiment, the aim is to establish a fixed electrostatic potential difference across a lipid membrane which contains a nanopore. Electrodes are used to make physical contact with the electrolyte solution in order to apply such a potential. If a randomly chosen piece of metal were used as an electrode in an electrolyte solution, the result of applying a potential would be that negative ions would migrate to the positive electrode, and positive ions would migrate to the negative electrode. But after a sufficient number of ions have accumulated near the electrodes so as to screen out the electric field, ions will cease to move, and ionic current flow will quickly decay away. Furthermore, Ohm's law (Eq. 2.2) makes it clear that the potential will also fall to zero. Only with the right electrode and electrolyte combination can an ionic current be maintained, and with it the electrostatic potential across the nanopore. If ions are to continue to flow indefinitely, there must be a source of ions at one electrode, and a sink for ions at the other electrode.

In most nanopore experiments, the electrode of choice is silver coated with silver chloride which is usually written as Ag/AgCl. It exhibits fast electrode kinetics so that the timescale of reactions at the electrode is many orders of magnitude faster than timescales that are experimentally accessible. Ag/AgCl electrodes are stable and simple to manufacture, the only requirements being a length of high-purity silver wire and some commercial laundry bleach. When the Ag wire is dipped into the bleach, the surface Ag atoms are oxidized, forming a coat of AgCl on the Ag wire. Also important is the fact that Ag/AgCl electrodes enable oxidation and reduction reactions involving chloride ions. The chloride ion is a practical choice for nanopore experiments because salts involving chloride are readily available and usually non-toxic. The following reactions take place at the surface of Ag/AgCl electrodes:

$$AgCl(s) + e^- \rightleftharpoons Ag(s) + Cl^- \tag{2.6}$$

In the forward reaction, which occurs at the negative electrode, the solid AgCl gets an extra electron from the external circuit, which reduces the silver, forming solid silver and a free chloride ion. In the reverse reaction, at the positive electrode, a chloride ion combines with solid silver, oxidizing it and forming AgCl, giving up a free electron to the electrode in the process. These oxidation and reduction reactions enable a steady ionic current to flow, which allows a potential difference to be maintained across a nanopore. In practice, however, the flow of ionic current can finally use up the thin film of AgCl on one of the electrodes and must be recharged. Most instruments incorporate AgCl pellets in the circuit which last much longer than a film on silver wires. Because silver ions interfere with the function of certain enzymes, the MinION uses a different electrochemical couple involving an iron compound called ferricyanide instead of silver. Ferricyanide is an iron atom surrounded by six cyanide molecules, and the iron can accept or donate electrons at electrodes just as silver does.

Resistance and Ionic Current in a Nanopore

The total resistance to current flow through a nanopore can be thought of as being the sum of two distinct contributions: (1) the resistance due to the geometry of the nanopore channel itself, R_p, and (2) the resistance due to the ions' ability to access the nanopore from bulk solution, called the "access resistance," R_a. This is depicted in Figure 2.2(a). The access resistance is given by[2,4]

$$R_a = \frac{\rho}{4r} \tag{2.7}$$

where r is the radius of the nanopore and ρ is the resistivity of the electrolyte solution whose standard international unit is Ohm-meter ($\Omega \cdot m$). If the nanopore itself is modeled as a conductive cylinder of length L and resistivity ρ, then the nanopore's resistance, R_p, is given by Eq. 2.1 and can be rewritten in terms of the nanopore radius as

$$R_p = \frac{\rho}{\pi r^2} L \tag{2.8}$$

The total resistance between the two electrodes is then

$$R_{\text{total}} = R_p + 2R_a = \frac{\rho}{\pi r^2} L + \frac{\rho}{2r}$$

$$= \frac{\rho}{\pi r^2}\left(L + \frac{\pi}{2}r\right) = \frac{\rho}{A}\left(L + \frac{\pi}{2}r\right) \quad (2.9)$$

When the access resistance is included, the total resistance is the same as that for a cylindrical resistor with a length $\left(L + \frac{\pi}{2}r\right)$ instead of L alone.[2] That amounts to approximately 1.6 nanopore radii, which is longer than typical nanopore channels used in sequencing.

It is straightforward to use Eq. 2.8 to estimate the total resistance of a nanopore in a membrane. For example if the inside radius of a nanopore's channel is 0.5 nm, and the length of this channel is 7 nm, then the total resistance in 1 M KCl would be on the order of $(0.1 \ \Omega{\cdot}m) \times (7 \ \text{nm} + 1.6 \times 0.5 \ \text{nm})/(\pi \times (0.5 \ \text{nm})^2)$, or about 1 G$\Omega$. At typical voltages used for nanopore experiments, around 100 mV, this would result in an ionic current flow of about 100 pA, or 10^{-10} A, a very small value. Since an ampere is 1 Coulomb/s, or 6.2×10^{18} ions/s, 100 pA amounts to approximately 600 million ions/s, or 600 ions/μs. This number is essential for nanopore sequencing. For instance, if it were only 6 ions/μs, a hundred times less, the modulations of ionic current caused by the four bases of a nucleic acid strand passing through the nanopore would be too small to be measured.

Capacitance in Nanopore Experiments

A biological lipid membrane acts as a capacitor. Imagine a membrane separating two chambers of electrolyte solution, without any nanopore, and no way for ions to pass from one side of the membrane to the other. When electrodes are used to apply a voltage across the membrane, there is a transient flow of ionic current. Negative ions flow away from the negative electrode and crowd up on one side of the membrane, while positive ions are pushed away from the positive electrode and crowd up on the opposite side of the membrane (Figure 2.2(b)). When the electrical potential difference across the lipid membrane equals the

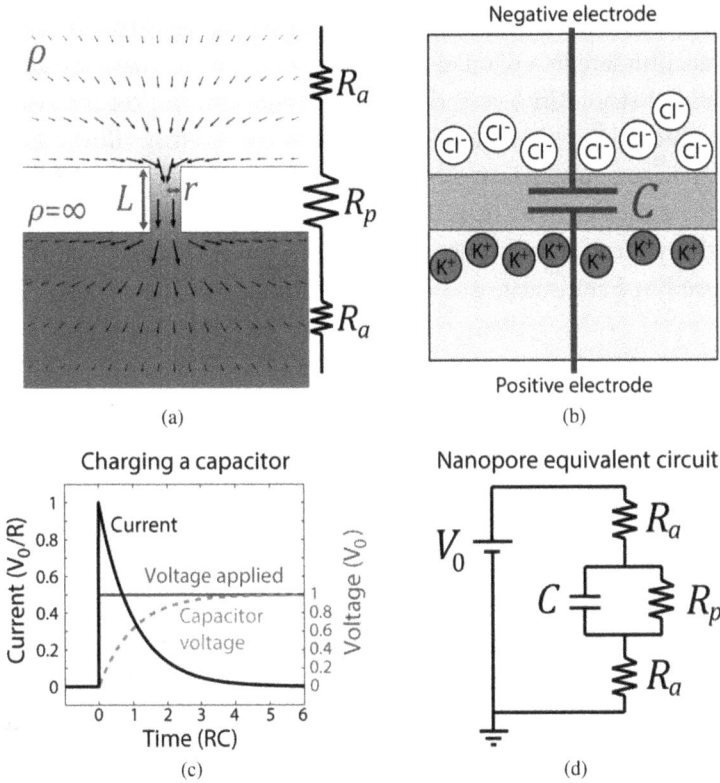

Figure 2.2. (a) Schematic of a nanopore modeled as a resistive cylinder in membrane of thickness L, shown here in cross-section. Arrows depict the path of ion flow. (b) The lipid membrane acts as a capacitor. When a voltage is applied across the membrane, ions line up on opposite sides, charging the capacitor. (c) Time course of the current flowing into a capacitor (solid black) in response to a voltage step (solid grey), as well as the time course of the potential across the membrane (dashed grey). (d) An equivalent electrical circuit for a nanopore experiment, accounting for the pore resistance, R_p, access resistance, R_a, and the membrane capacitance, C.

voltage applied at the electrodes, the transient ionic current ceases. Now imagine instead that a nanopore is present in the lipid membrane (as was shown in Figure 2.2(a)). An initial transient flow of ions still occurs, during which ions crowd up on both sides of the membrane, but when this transient ionic current subsides, there remains a steady-state current determined by the nanopore's resistance to ionic flow

(Eq. 2.9). The initial transient current flow is similar to the way electrons accumulate in a parallel-plate capacitor, except that here opposite charges are stored in a very thin layer of the solution on the two sides of the lipid bilayer that significantly resists charge from flowing.

Capacitors are characterized by a value called the capacitance, C, which is a measure of their ability to store a given amount of charge, Q, with a potential difference, V, between the two plates. Capacitance, measured in Farads (F), is defined such that

$$C \equiv Q/V \qquad (2.10)$$

In the case of an ideal parallel-plate capacitor, the capacitance can also be written as

$$C = \frac{\epsilon A}{d} \qquad (2.11)$$

where ϵ is the dielectric constant of the material separating the plates, A is the area of a plate, and d is the separation between them. We can see that for a lipid bilayer, where d is the thickness of the bilayer (only a few nanometers), capacitance can be relatively large, depending on the area of the bilayer. As an order of magnitude estimate, supposing a bilayer area of 500 μm^2 and a membrane hydrophobic thickness of 4 nm, and using $\epsilon \approx 1.9 \times 10^{-11}$ F/m for hydrocarbons,[5] the expected capacitance would be around 2 pF.

At equilibrium, a capacitor with capacitance C will hold a charge Q when a voltage V is applied. When the voltage is first applied, a current must flow in order to bring charge onto the capacitor. The initial current flow due to a capacitor in the circuit is temporary, since once the capacitor attains a potential V across its plates, no more charge will flow. This capacitive current is described by the equation

$$I = C \frac{dV}{dt} \qquad (2.12)$$

When in series with a resistor, such as an electrolyte solution, the membrane becomes part of what is called an "RC circuit" and the quantity RC has units of time, and is the time constant for charging the membrane.

A plot of the membrane voltage and the measured current in response to an applied voltage step is shown in Figure 2.2(c). Thus, when a voltage potential is turned on, or changed in magnitude, the applied potential will show up across the membrane with a time delay. The larger the capacitance, the more slowly an applied voltage step will reach the membrane. For a typical lipid bilayer membrane, the time constant for an applied potential to reach the nanopore would be around 10 ns. Using 100 mM KCl as the electrolyte and increasing the bilayer area by a factor of ten can lead to timescales on the order of microseconds.

Putting everything together, the nanopore setup looks like Figure 2.2(d) from an electrical standpoint. The capacitance of the membrane is in parallel with the nanopore resistance. The access resistance combines in series with the membrane capacitance to filter out high-frequency changes in membrane potential, and the total resistance of the circuit is a sum of the nanopore resistance and the access resistance. The goal of a nanopore sequencing experiment is to measure the small current which flows through this circuit when a potential is applied.

Measuring Small Currents

Small currents can be measured using operational amplifiers [op-amps], as shown schematically in Figure 2.3. The function of an op-amp in the

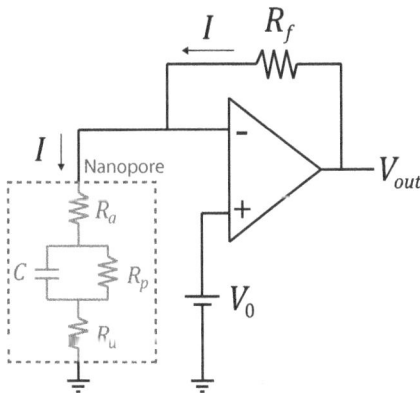

Figure 2.3. An operational amplifier used in a current amplifier configuration. Commercial instruments capable of measuring picoampere currents are more sophisticated realizations of this general principle.

configuration shown is to force the negative terminal to match the voltage V_0, applied at the positive terminal, as quickly as possible. The resulting current, I, that flows through the nanopore equivalent circuit must also flow through the feedback resistor, R_f since the op-amp's terminals draw no current. Thus the voltage at the op-amp's output is $V_{out} = IR_f$. Although I is on the order of picoamperes, the choice of a very large feedback resistor, say 1 GΩ, means that the output voltage is on the order of millivolts, a quantity large enough to measure. This output voltage can be used to deduce the current through the nanopore, $I = V_{out}/R_f$.

Noise

Attempts to measure any small signal will quickly reveal that small signals are often obscured by noise. Elimination of as many noise sources as possible is a necessary precursor to nanopore sequencing. Noise takes on several forms, each caused by a different underlying physical mechanism. Central to an understanding of noise is the phenomenon of thermal motion, also called Brownian motion. Due to the equipartition of energy in statistical physics, particles at a finite temperature T have an associated kinetic energy of $\frac{1}{2}k_BT$ per degree of freedom, where k_B is Boltzmann's constant. This means that particles, such as ions or water molecules for example, have a translational kinetic energy $\frac{1}{2}mv^2 = \frac{3}{2}k_BT$, and so the root mean square velocity of a particle with mass m is $v = \sqrt{\frac{3k_BT}{m}}$. For a potassium ion at 20°C, this comes to 433 m/s, an incredible speed. At that speed, the ion is constantly colliding with neighboring molecules, and it is helpful to keep this in mind when imagining what happens down near the length scale of single molecules. This sort of random, rapid motion is everywhere on the molecular scale, and it is often referred to as thermal motion.

Thermal motion of charge carriers in a resistor leads to small voltage fluctuations, since each charge carrier has an electrostatic potential associated with it. When the individual charges move, the local potential changes slightly. Thermally induced voltage fluctuations called noise occur in all resistors, giving rise to tiny fluctuating currents, even without a voltage bias applied. These fluctuations are called Johnson–Nyquist noise. This type of noise is independent of frequency. Noise is usually measured in terms of the current noise power spectral density (PSD), $S_1(f)$ PSD is the frequency response of a random or periodic signal. It tells us where the average power is distributed as a function of frequency. For Johnson–Nyquist noise, $S_1(f) = 4k_B T/R$. A simulated plot of Johnson–Nyquist noise is shown in Figure 2.4(a), using

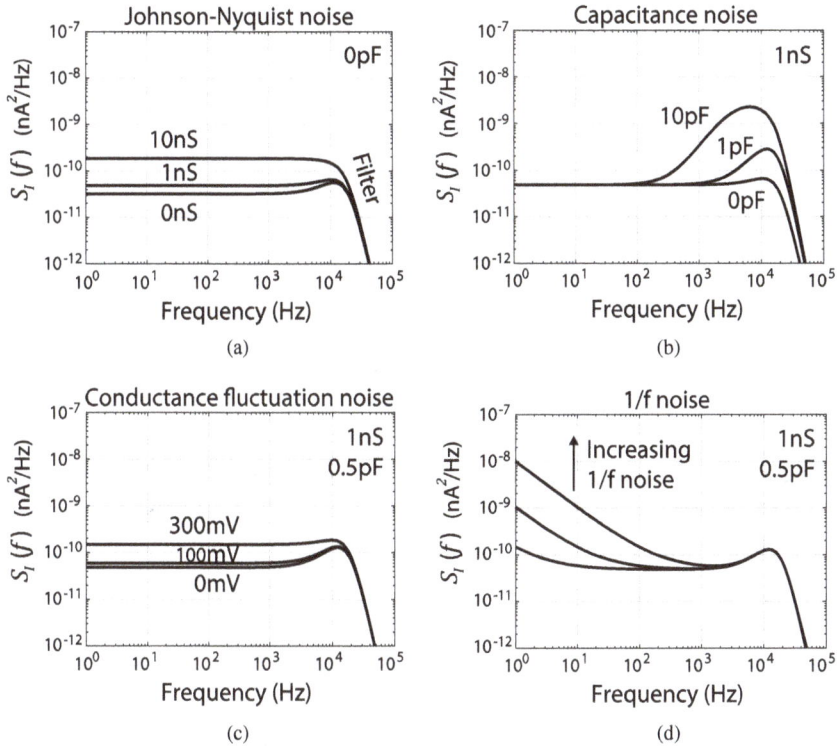

Figure 2.4. (a) Idealized current noise PSD, $S_1(f)$, for Johnson–Nyquist noise, plotted for different nanopore conductance values, using parameters typical for a commercial current amplifier. The slight rise in noise near 10 kHz is caused by capacitance at the current amplifier's input. The effect of a 10 kHz filter is shown as the drop-off in noise around that frequency. (b) Idealized current noise PSD for capacitance noise, plotted for a 1 nS nanopore in a membrane with different capacitances. (c) Idealized current noise PSD for conductance fluctuations, plotted for at various applied voltage biases. (d) Idealized current noise PSD for a 1 nS nanopore exhibiting different amounts of 1/f noise. Fortunately, 1/f noise is usually not detected in protein nanopores, and its presence indicates that something about the nanopore setup has gone wrong.

typical values for nanopore resistance, op-amp feedback resistance, and temperature.

When the same voltage fluctuations occur across a capacitor, they lead to capacitance noise. The associated current noise PSD is proportional to the square of capacitance as well as the square of the frequency. When attempting to measure current on short timescales, capacitance noise becomes increasingly problematic, and this type of noise is usually subdued by using a low-pass filter to remove signal and noise at the high frequencies. Capacitance noise is plotted in Figure 2.4(b).

Voltage fluctuations are not the only physical mechanism that cause current noise. The number of ions in the nanopore is not completely constant as a function of time. Fluctuations in the number of charge carriers in the nanopore lead to fluctuations in the channel conductance, and thus the current. These fluctuations give rise to a current noise PSD independent of frequency and proportional to the square of the current which flows through the nanopore[6]: $S_I(f) \propto I^2$. This type of noise is depicted in Figure 2.4(c), using estimated values similar to those observed experimentally.

Perhaps the most important and immediately apparent source of noise is from the world outside the experimental apparatus. Any stray electric or magnetic field can induce currents to flow. In the case of ubiquitous line noise (60 Hz or 50 Hz), the induced currents can be far larger than the nanopore signals of interest. This sort of extraneous noise must be eliminated. A Faraday cage, which is nothing more than a highly conductive metal enclosure, is typically employed for this purpose. Faraday cages screen out stray electric fields. The free electrons at the outer edges of a highly conductive metal respond very quickly to external electric fields by moving, and they continue to move until they cancel out the external field — if they did not cancel the field, they would continue to experience a force. This rearrangement of charges leads to an interesting and useful property of conductors: the electric field inside an ideal conductor, or conductive shell, is zero. A Faraday cage uses this principle to achieve electric field cancellation within a metal box, where the experiment is housed. Proper implementation reduces externally induced current noise to levels below that of Johnson – Nyquist noise. Current noise induced by external sources can be identified as one or more sharp spikes in the PSD.

A final and rather mysterious type of noise is called $1/f$ noise. The name arises from the PSD of $1/f$ noise, where $S_1(f) \propto 1/f$. The large amount of current noise at low frequencies manifests as a slow drift in the mean ionic current. This type of noise is very problematic when it shows up in nanopore experiments, since nanopore sequencing relies on being able to accurately and reproducibly measure mean ionic current levels. And averaging a signal over a longer period of time does not help, since the mean drifts more! $1/f$ noise is ubiquitous in nature, though its physical basis can be different in different circumstances, and is only known in some specific cases. The current noise PSD characteristic of $1/f$ noise is shown in Figure 2.4(d). Fortunately, $1/f$ noise does not usually appear in α-hemolysin, MspA, or CsgG nanopores in lipid membranes. If it does show up in nanopore sequencing experiments using protein pores in lipid membranes, it indicates the presence of contaminants or a defect in the membrane itself.

Limits on Temporal Resolution of Ionic Current Measurements

During free translocation with an applied potential difference of about 100 mV across the membrane, ssDNA transits a nanopore at speeds on the order of 1 μs per base. Measuring the small currents that flow through a nanopore during only 1 μs poses several major problems.

Typical currents as the ssDNA transits a nanopore are on the order of 10 pA, which would translate to an average of only about 60 ions/μs. Even if the current measurement were subject to no other noise sources, the precision in the measure of the number of ions that flow will be the standard deviation of that number divided by the mean. Because we must measure the differences between one DNA base and its neighboring base, and because only about 60 ions/μs will be the current during occupancy by any base, the measurement precision could not be better than about $(60)^{\frac{1}{2}}/(60) \sim 0.13$ or $\pm 13\%$. Because differences between currents associated with different nucleotides can be on the order of 1 pA or less, the current measurement precision must be considerably better than $\pm 13\%$ to distinguish these small differences. Fast current measurement also runs into a problem due to capacitance noise. Because capacitance noise is proportional to the square of the frequency, this will overwhelm signals on short time-scales. Furthermore, op-amps also have inherent limitations in their high frequency response. All of these factors impose a fundamental physical limitation which can only be overcome by slowing the rate of DNA translocation through the nanopore, i.e. increasing the length of time each nucleobase remains in the nanopore.

To date, the only successful strategy to slow ssDNA translocation has been to use of a processive enzyme to control the nucleobases' movements through a nanopore. As explained in Chapter 6, motor enzymes such as polymerases or helicases can bind to ssDNA and ratchet the DNA through the nanopore in discrete steps that are slow enough to permit accurate current measurements during each step.

References

1. Weast, R.C. & Lide, D.R. *CRC Handbook of Chemistry and Physics*, Edn. 70 (CRC Press, Boca Raton, FL; 1989).
2. Hille, B. *Ion Channels of Excitable Membranes*, Edn. 3 (Sinauer Associates, Sunderland, MA; 2001).
3. Bard, A.J. & Faulkner, L.R. *Electrochemical Methods: Fundamentals and Applications*, Edn. 2 (John Wiley & Sons, New York; 2001).
4. Hall, J.E. Access resistance of a small circular pore. *J. Gen. Physiol.* **66**, 531–532 (1975).
5. Montal, M. & Mueller, P. Formation of bimolecular membranes from lipid monolayers and a study of their electrical properties. *Proc. Natl. Acad. Sci. USA* **69**, 3561–3566 (1972).
6. Mak, D.O. & Webb, W.W. Conductivity noise in transmembrane ion channels due to ion concentration fluctuations via diffusion. *Biophys. J.* **72**, 1153–1164 (1997).

Chapter 3

Lipids and Nanopores

David W. Deamer

Nanopore sequencing depends on the fact that certain protein channels can be inserted into a membrane that separates two compartments. For this reason, it is important to understand the physical and chemical properties of the specialized molecules that compose the membrane. Lipids are generally defined by the fact that one portion of a lipid molecule is composed of hydrocarbons. By themselves, hydrocarbons are oils that are soluble in organic solvents such as chloroform, ethanol, or diethyl ether. Because of their oily hydrocarbon composition, lipids are also soluble in organic solvents. Hydrocarbons are insoluble in water and are referred to as being hydrophobic.

Virtually all lipids also have a hydrophilic group such as carboxylate, phosphate, or hydroxyl attached to one end of the hydrocarbon portion. These groups are polar or ionic and therefore interact strongly with water. The combination of a hydrophobic hydrocarbon and a hydrophilic group on the same molecule is the reason why lipids are called amphiphiles, a term derived from Greek words meaning 'loving both'. Their amphiphilic character gives lipids unique physical properties that are essential to all life, and also to nanopore sequencing.

Lipid Molecules

The simplest compounds that qualify as lipids are fatty acids composed of a hydrocarbon chain with a carboxyl group at the end. Some

Fatty Acid Structure

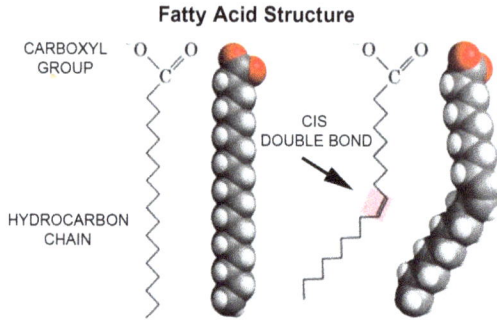

Figure 3.1.   Fatty acids are composed of a hydrocarbon chain with a carboxyl group at one end. Some fatty acids are unsaturated, which means that they have one or more *cis* double bonds in the hydrocarbon chain. Stearic acid (left) and oleic acid (right) are shown here, each with 18 carbon atoms.

fatty acids are saturated and have straight chains while others have one or more double bonds (Figure 3.1). The double bond puts a kink into the chain, and this prevents the hydrocarbon chains from packing tightly, thereby increasing the fluidity that is essential for membranes to function in cells. Although fatty acids can assemble into membranes under certain conditions, the membranes are relatively unstable, so no living organism uses fatty acids as primary membrane components.

At some point early in evolution, microbial life evolved biochemical pathways that attach two fatty acids to a glycerol phosphate. Having two chains on the same molecule markedly increases membrane stability, and the resulting phospholipids became the basic amphiphilic compound present in all cellular membranes.

Figure 3.2 shows several species of phospholipids that are components of biological membranes. Phosphatidic acid is present only as a metabolic intermediate that is converted to other phospholipids when a second group such as choline, ethanolamine, serine, or glycerol is attached to the phosphate. The fatty acids of phospholipids are typically in the range of 14–18 carbons in length. Shorter chains are unable to assemble into stable membranes, while longer chains can potentially "freeze" into a relatively immobile gel state. As will be discussed below, all membranes must be in a fluid state in order to function.

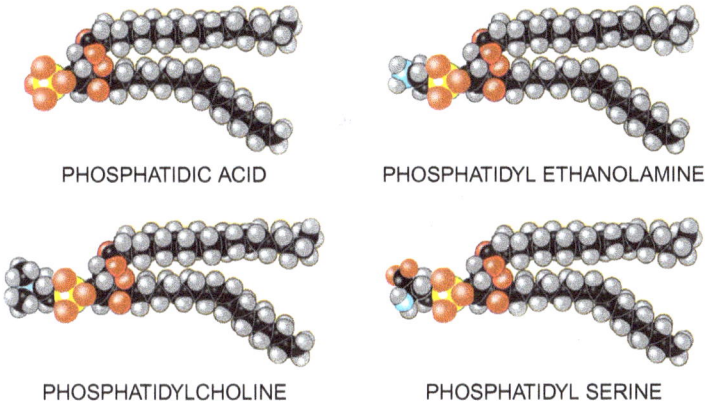

Figure 3.2. Structures of common phospholipids. Color code for atoms: Black = carbon; white = hydrogen; red = oxygen; blue = nitrogen; yellow = phosphorus.

Phospholipids typically have one saturated and one unsaturated fatty acid that are linked to the glycerol through ester bonds. Stearic acid, with 18 carbons, is an example of a saturated fatty acid, and oleic acid also has 18 carbons but with a *cis* unsaturated bond between the 9 and 10 carbons. The *cis* double bond puts a kink into the chain that significantly lowers its melting point. For instance, the melting point of stearic acid is 68°C, while oleic acid melts at 14°C.

Although most phospholipids have fatty acids attached through ester bonds, a few lipids, particularly extremophilic microorganisms like the Archaea, use ether linkages which are much more stable to hydrolysis. Another accommodation to high temperatures and extreme pH ranges is hydrocarbon chains with several methyl groups attached instead of unsaturated bonds, which are readily damaged by reacting with molecular oxygen in air. The branched chains serve the same purpose as unsaturation by increasing the fluidity of the membranes that would otherwise freeze into a gel state. Because the lipid membranes supporting nanopores must work in air, in order to avoid oxidation damage, a branched chain lipid called diphytanoyl phosphatidylcholine is often used to support protein nanopores.

Self-assembly of Amphiphiles

All lipids are surface active and would be classified as surfactants, a property shared with commercial detergents used as cleaning agents. When added to water, surfactants accumulate at the air–water interface to form a monomolecular layer, the simplest example of which is called self-assembly. The hydrophobic portion of a surfactant is insoluble in water, and the hydrophilic group interacts strongly with water, with the result that the hydrophobic groups form a continuous oily layer between the water and air, but are anchored there by the hydrophilic groups. However, the amphiphilic molecules are not exclusively at the surface, but may also disperse in the water phase as micelles and liposomes. Micelles have a purely hydrophobic interior composed of hydrocarbon chains, while liposomes are defined by a lipid bilayer boundary that contains an interior aqueous volume (Figure 3.3).

Amphiphilic compounds can also assemble into planar bilayers. In the 1960s, several investigators began to study model membrane systems composed of planar bilayers. One of the methods, pioneered

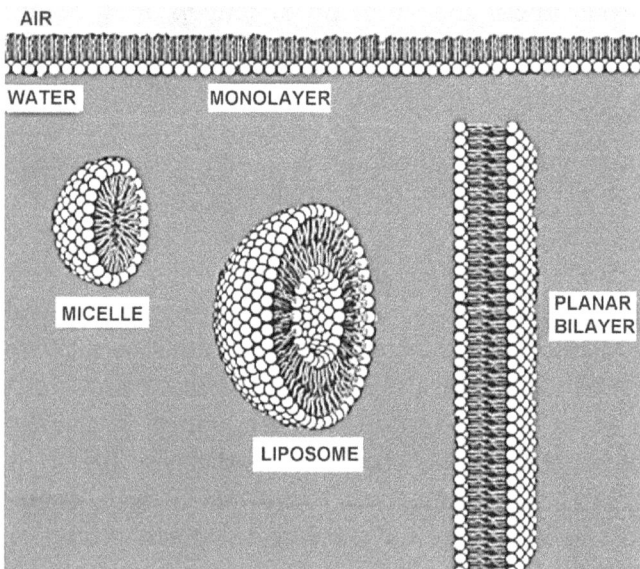

Figure 3.3. Self-assembled structures of amphiphilic molecules in water.

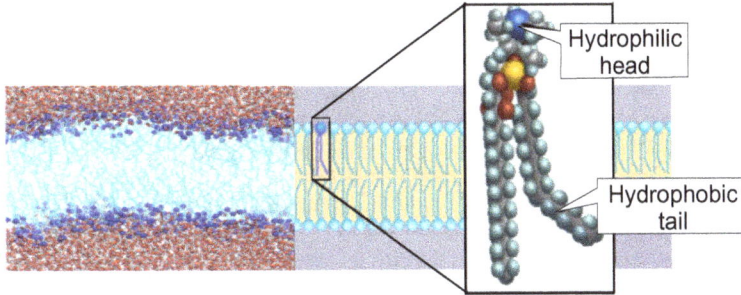

Figure 3.4. The left section shows a computer simulation of a phospholipid bilayer in water which is usually depicted in the simplified version shown on the right. Image courtesy of Andrew Pohorille.

by Mueller and Rudin[1] involved "painting" a mixture of phospholipid and an alkane such as tetradecane or hexadecane across a small aperture in the surface of a thin sheet of a hydrophobic plastic like Teflon. The mixture slowly thins to become a lipid bilayer supported by a torus of the oil–lipid mixture around the edge of the aperture. At that point the bilayer does not reflect light so it was originally referred to as a "black lipid membrane" or BLM, but the term planar lipid bilayer is more commonly used today. These early studies provided the foundation for later work in which lipid bilayers were used to support protein nanopores that allowed DNA molecules to be translocated through the membrane.

Figure 3.4 shows a computer simulation of a planar phospholipid bilayer in water. Hydrophilic head groups (dark blue) are interacting with water molecules (red), and hydrophobic hydrocarbon chains (light blue) are in the interior. They look disordered, and this is because the phospholipid molecules are in constant diffusive motion in the plane of the membrane and their hydrophobic tails are melted and also in constant fluid motion. Certain protein channels can insert spontaneously into such fluid planar bilayers and provided the first nanopore system shown to detect nucleic acids.

The last amphiphilic compounds to be considered are block copolymers. Certain microorganisms called hyperthermophiles thrive in near boiling water temperatures of hot springs. Lipid bilayers would

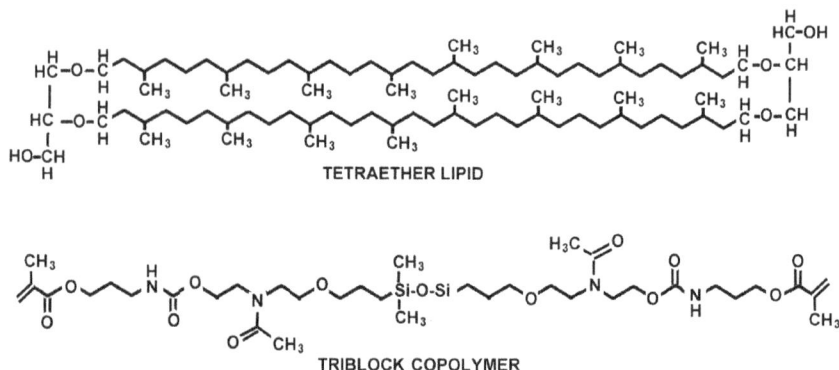

Figure 3.5. Comparison of a tetraether lipid with a triblock copolymer.

not be sufficiently stable at those temperature ranges, so hyperthermophiles synthesize tetraether lipids, an example of which is shown in Figure 3.5. The hydrocarbon chains are long enough to span the membrane and are therefore present as a monolayer in the membrane, rather than a bilayer. The molecules are highly branched with multiple methyl groups ($-CH_3$) along the chains, and ether linkages are used instead of ester bonds. The hydrophilic ends are simply glycerol with its hydroxyl group rather than phosphate.

Tetraether lipids would have little relevance to nanopore sequencing except that nanopore membranes face the same stability problems as hyperthermophiles. The commercial devices must stand up to being shipped to users worldwide, so fragile planar bilayers would not stand up to this kind of stress. The solution was to use a synthetic version of monolayer membranes based on block copolymers. Although the commercial membrane component is proprietary, a published block copolymer that assembles into a very stable membrane is also shown in Figure 3.5.

Forces Stabilizing Self-assembly of Amphiphiles

Why can certain lipids spontaneously assemble into bilayer structures? There are two primary forces that stabilize their assembly. The first is an intermolecular force called van der Waals interactions, named after the 19th century scientist who first described them. The electron shells of molecules are not fixed around the nuclei of the atoms that compose them, but instead have fluctuations that cause slight transient differences in the electrical charge of the surface. When two molecules are near each other, the fluctuating charges on one surface induce opposite charges in the second surface, producing an attractive force. Although van der Waals interactions are much weaker than ionic and covalent bonds between atoms, in aggregate the forces can have a significant effect on the physical properties of molecules. For example, methane, ethane, propane, and butane with 1, 2, 3, and 4 carbons are all gases at ordinary temperatures and pressure, but a pentane molecule, with 5 carbons, has sufficient van der Waals interactions with neighboring pentane molecules to be a liquid at room temperature. All saturated hydrocarbons between pentadecane and hexadecane are liquids at room temperature, but longer hydrocarbons are solids. The same forces act between lipid molecules in a bilayer, and are therefore important for nanopore sequencing because they act to stabilize the bilayer.

The second physical effect that stabilizes lipid bilayers is more complicated. Imagine that a hydrocarbon chain similar to one of the two hydrocarbon tails of a phospholipid molecule like phosphatidylethanolamine (Figure 3.2) could somehow be forced into water. This would require energy for two reasons. First, in order to insert a hydrocarbon chain into water, the chain must disrupt hydrogen bonds between water molecules, which means that energy is being added to the system. Second, the presence of the hydrocarbon chain causes water molecules to form an ordered layer of water around the chain. If one chain happens to encounter another hydrocarbon chain, bringing the two chains together, some of the hydrogen bonds between water molecules can form again. Furthermore, the organized water molecules from around each of the chains are dispersed back into the surrounding medium, which allows them to become disordered. A principle of thermodynamics is that any orderly system of molecules tends to become more disordered. In other words, if a system can gain entropy (entropy is a measure of disorder), it will do so. Because water molecules are small, a large number of waters can potentially come into contact with the hydrocarbon tails of a single phospholipid. Thus, the system as a whole gains an overwhelming amount of entropy simply by bringing the hydrocarbons tails together and excluding many water molecules from contact with the lipid's hydrocarbon tails where those molecules would otherwise form a well-ordered layer. The process that brings together hydrophobic components of molecules and excludes water from contact with these components is sometimes called the hydrophobic effect, or entropic bonding. It explains why, as the old saying goes, water and oil don't mix. Because the hydrocarbon chains of phospholipids experience the hydrophobic effect, they spend most of the time interacting with other hydrocarbon chains in the bilayer interior. Thus, the combination of many weak van der Waals interactions between lipid molecules and the hydrophobic effect that excludes water from interaction with phospholipid tails strongly stabilize a bilayer.

Properties of Lipid Bilayers: Permeability

The simplest function of the lipid bilayer in a biological membrane is that it provides a permeability barrier limiting free diffusion of ionic and polar solutes through the membrane. Although such barriers are essential for cellular life to exist, there must also be a mechanism by which selective permeation allows specific solutes to cross the membrane. In contemporary cells, transport of ions and nutrients are mediated by transmembrane proteins that act as channels and transporters. Examples include the proteins that facilitate the transport of glucose and amino acids into the cell, channels that allow potassium and sodium ions to permeate the membrane, and active transport of ions by enzymes that use ATP as an energy source.

To give a perspective on permeability and transport rates by diffusion, we can compare the fluxes of relatively permeable and relatively impermeable solutes through lipid bilayers. A quantitative measure of permeability is given by a parameter called the permeability coefficient (P), which is determined by measuring the flux of a solute across a unit area of membrane driven by a concentration gradient. A simple version of the equation defining the permeability coefficient

(P) is $P = J/\Delta C$, where J is the measured flux expressed as moles of solute cm^{-2} s^{-1}, and ΔC is the difference in concentration of the solute across the membrane with units of moles cm^{-3}. The units of P are expressed as centimeters per second (cm s^{-1}). To give some examples, the permeability coefficient of water through a lipid bilayer is approximately 10^{-3} cm s^{-1}, and the permeability coefficient of potassium ions is 10^{-11} cm s^{-1}. It follows that the measured permeability of lipid bilayers to small, uncharged molecules such as water is greater than the permeability to ions by a factor of ~10^8. These values mean little by themselves, but make more sense when put in the context of time required for exchange across a bilayer: half the water in a liposome exchanges in a few milliseconds, while potassium ions have half-times of exchange measured in days.

The reason that potassium ions are so impermeable arises from an effect called Born energy, named after Max Born, a German physicist who won the Nobel prize for his research on quantum theory. Born pointed out that ions have a property he called self-energy which is associated with their electric field. As a result of the field, ions in a polar medium like water strongly interact with the electrical charges on the water molecules, and those interactions must be broken to move the ion into a nonpolar medium like the oily phase of a lipid bilayer. This requires a very large expenditure of energy, one estimate being ~40 kcal $mole^{-1}$, so virtually no ions are able to penetrate the bilayer. Selective permeability to ions in the membranes of living cells is provided by specialized transmembrane proteins that produce a water-filled channel through the hydrocarbon layer. Early research on nanopore sequencing depended on one such channel called alpha hemolysin that allows ions and DNA to flow through when a voltage is applied.

Fluidity

In the 1960s, our increasing understanding of membrane structure began to make it clear that the proteins of biological membranes are embedded in a fluid sea of lipids. The first indication that membranes were surprisingly fluid was an experiment by Frye and Ededin[2] who

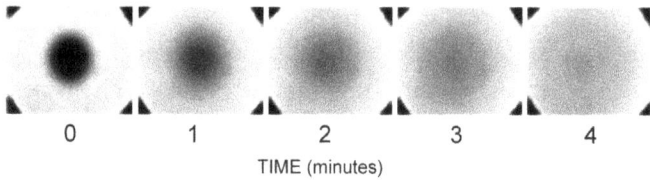

Figure 3.6. Demonstration of membrane fluidity. Depicted here is a lipid bilayer supported across a small aperture. The lipid has a fluorescent label, but when it is hit by a focused beam of laser light, the fluorescence is bleached to a dark spot. Within a few minutes, the spot disappears as fluorescent lipids diffuse into the bleached region from the surrounding unbleached areas of the bilayer. Adapted from Ref. 3.

used fluorescent dyes to label the membranes of cells, then caused the cells to fuse. They observed that the labeled components of two separate membranes began to diffuse into each other within minutes, and concluded that the membrane had a fluid character rather than being fixed in place. The fluidity of lipid bilayers has since been clearly demonstrated by using focused laser light beam to bleach a small area of a fluorescent lipid bilayer (Figure 3.6). Within seconds, the diffusion of lipids within the bilayer begins to fill in the bleached spot, which completely disappears 4 min later.

Based on this and other observations, the fluid mosaic concept was proposed by Singer and Nicolson[4] as a plausible model to account for the properties of biological membranes. Since that time, numerous studies have measured the diffusion coefficient of lipids and proteins in membranes, and the diffusion rates were found to correspond to those expected of a fluid with the viscosity of oil, rather than a gel phase resembling wax. The fluidity of the lipid phase has ramifications in nanopore sequencing because the protein nanopore is not fixed in place but instead is diffusing freely in the plane of the bilayer into which it is inserted. This can limit the lifetime of the pore because the protein channel can move to the rim of the aperture supporting the bilayer and adhere to it in such a way that it can no longer conduct ionic current.

Capacitance and Noise

A common experience is to be listening to a radio program from a distant station. Sometimes the signal is clear and understandable,

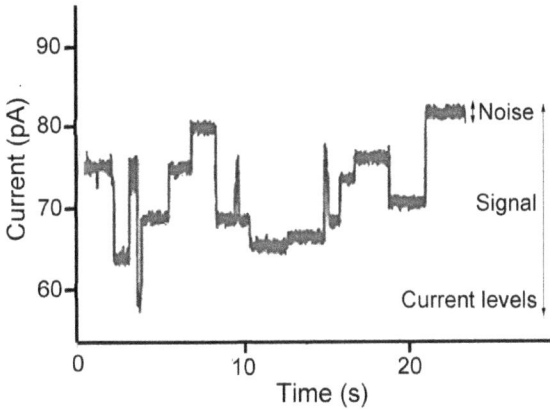

Figure 3.7. Signal-to-noise ratio in a nanopore measurement. Each jump in current level is due to a single base in a DNA strand being advanced through the nanopore, whereas the noise is related to the signal amplification system and the capacitance of the lipid bilayer.

but then is drowned out by the random noise of static. Virtually all measurements of physical processes in the laboratory are also a combination of signal and noise. In the specific case of nanopore sequencing, the signal is produced by a strand of DNA moving through the pore, and the noise is the random background of variations in the electrical signal that is partially due to the instrument itself, but also arises from the membrane and the way that ions move through the pore (Figure 3.7). Just as we prefer a clear signal free of static when listening to the radio, it is also important to minimize the noise when performing nanopore sequencing. A detailed treatment of noise related to nanopore sequencing is described in Chapter 2.

Bilayer Stability

Ideally, the lipid membranes supporting a protein nanopore would be indefinitely long-lived. However, a membrane just 5 nm thick is inherently fragile and sooner or later breaks, bringing an end to its useful life. From practical experience, it is known that several factors limit the stability of bilayers, and these are worth understanding in order to maintain a functioning membrane as long as possible.

Contamination by other components in the system is the most common cause of bilayer disruption. For instance, nanopore analysis requires high concentrations of ionic solutions such as 1.0 M KCl in order to provide the ionic current required to monitor translocation of a DNA strand through the pore. It is impossible for reagents to be completely pure, but if care is taken to purchase reagent-grade salts and buffers, their solutions are likely to be pure enough that they do not cause membranes to break unless the water used to make the 1.0 M KCL contains contaminants such as traces of detergent-like molecules.

Another important source of contaminants may be present in the lipids used to form the membrane. Lipids can undergo decomposition reactions such as oxidation of unsaturated bonds and hydrolysis of ester bonds linking fatty acids to the glycerol in phospholipids. In both cases, the degradation products have detergent-like properties. They accumulate in the membranes, first producing undesirable channel-like conductance and finally causing the membrane to break. For this reason, only the purest lipids should be used to prepare bilayers and the stocks should be kept frozen and away from air to prevent oxidation.

The aperture used to support the lipid bilayer is typically composed of an inert material such as Teflon. Over time and during use, the surface inevitably accumulates a coat of lipids and oils that can affect the stability of lipid bilayer. It is essential for the support to be vigorously cleaned between uses, for instance, by boiling for a few minutes in 10% nitric acid. Because bare skin also is coated in a thin layer of oil, protective gloves should always be worn while handling the components of a nanopore system.

Vibration can also cause instability of bilayers and add noise to measurements. For this reason, nanopore devices are usually installed on specialized tables that dampen vibration with air cushions that support the surface. All of the above advice is applicable only to ordinary nanopore devices assembled in the laboratory for research using single channels such as hemolysin. It does not apply to the MinION, which has 2,000 pores in membranes that are so stable they can stand up to being shipped as air freight. However, over many hours, even the nanopores in the MinION device are lost one by one and a new flow cell will eventually be required.

Insertion of Peptides and Proteins into Lipid Bilayers

The first artificial ion channel was discovered by Hladky and Hayden[5] who found that an antibiotic called gramicidin could assemble within a lipid bilayer to form cation-selective channels that allowed sodium and potassium ions to diffuse across the membrane. Their discovery showed that it would be possible to experimentally model the ion-selective channels that are present in the membranes of all living cells. Over the next decade, it turned out that gramicidin was one of many antibiotic molecules produced by microorganisms that could alter the permeability of lipid bilayers to ions. Other examples include valinomycin, nigericin, and alamethicin, all of which are small peptides. Gramicidin, for instance, has just 15 amino acids that turn into a partial helical barrel in the hydrophobic interior of a lipid bilayer. A single gramicidin is not large enough to be a channel by itself, but when two gramicidin molecules happen to meet, they form a transient head-to-head hydrogen bond lasting for a few seconds (Figure 3.8). This briefly stabilizes their combined helical structures which spans the bilayer in the form of a tunnel-like barrel that accommodates a strand of ~6 water molecules. Ions like sodium or potassium can pass through the tunnel, pushing the water molecules ahead of them.

Certain proteins can also assemble into much larger transmembrane channels that conduct ions. In 1970, an antibiotic protein was reported to be produced by *Staphylococcus* bacteria and named α-hemolysin.[6] Over the ensuing years, interest in the pore-forming ability of α-hemolysin slowly increased, and in 1987 it was demonstrated to form pores in lipid bilayers.[7] This explained its cytotoxic effect and in particular why it causes erythrocytes to swell and release hemoglobin, a process called hemolysis from which hemolysin gets its name. The protein channel used in the MinION is called CsgG. It is not an antibiotic, but instead is a porin isolated from the membranes of *E. coli* bacteria. The next chapter will describe the properties of

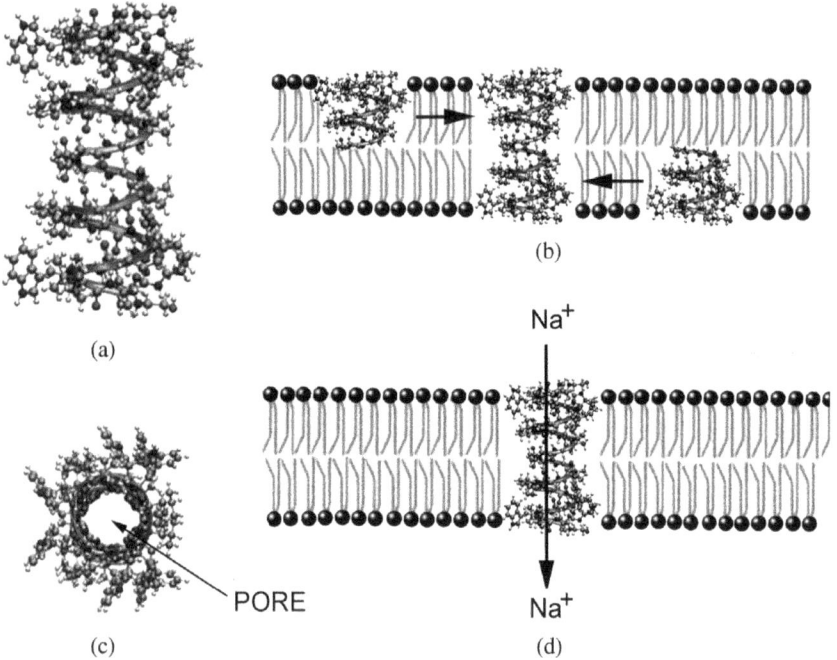

Figure 3.8. Gramicidin is one of the simplest channel-forming antibiotic peptides. Each molecule has just 15 amino acids and can spontaneously become embedded in a fluid lipid bilayer. When two gramicidin molecules happen to meet, they form a transient transmembrane channel that conducts ionic current for a few seconds before falling apart.

these specialized proteins that have been used to develop nanopore sequencing of DNA.

References

1. Mueller, P. & Rudin, D.O. Action potentials induced in biomolecular lipid membranes. *Nature* **217**, 713–719 (1968).
2. Frye, L.D. & Edidin, M. The rapid intermixing of cell surface antigens after formation of mouse-human heterokaryons. *J. Cell Sci.* **7**, 319–335 (1970).
3. Cho, N.-J., Hwang, L.Y., Solandt, J.J.R. & Frank, C.W. Comparison of extruded and sonicated vesicles for planar bilayer self-assembly. *Materials* **6**, 3294–3308 (2013).

4. Singer, S.J. & Nicolson, G.L. The fluid mosaic model of the structure of cell membranes. *Science* **175**, 720–731 (1972).
5. Hladky, S.B. & Haydon, D.A. Ion transfer across lipid membranes in the presence of gramicidin A: I. Studies of the unit conductance channel. *Biochim. Biophys. Acta* **274**, 294–312 (1972).
6. Sengers, R.C.A. Hemolytic action of staphylococcal α-hemolysin on human erythrocytes in a Na^+- and K^+-containing suspending fluid. *Antonie van Leeuwenhoek* **36**, 57–65 (1970).
7. Belmonte, G. *et al.* Pore formation by *Staphylococcus aureous* alpha-toxin in lipid bilayers: Dependence upon temperature and toxin concentration. *Eur. Biophys. J.* **14**, 349–358 (1987).

Chapter 4

Nanopore Structure, Assembly, and Sensing

Daniel Branton

The "nanopores" used in nanopore sequencing are bacterial proteins that are able to spontaneously insert into cellular or artificial lipid-bilayer membranes. Upon insertion, they form an aqueous channel that can conduct ions through the membrane. When sequencing DNA or RNA, such nanopores serve as the primary sensor that distinguishes and identifies the sequence of nucleotides in a nucleic acid polymer. Remarkably, none of the proteins that have been used in nanopore sequencing (Figure 4.1) evolved in nature as sensors or detectors of nucleobases. They evolved to perform a variety of bacterial functions such as nutrient import, secretion, or as antibiotics, and it is astonishing that these otherwise unrelated proteins are such effective nucleobase sensors. What are the common features that allow them to identify the purines and pyrimidines in polymers of DNA and RNA? Because the α-hemolysin channel was the first to demonstrate the potential of nanopore sequencing, it has been extensively investigated and its properties will be the initial focus of this chapter. We will then use those properties as a way to understand why other channel proteins are now preferred for nanopore sequencing.

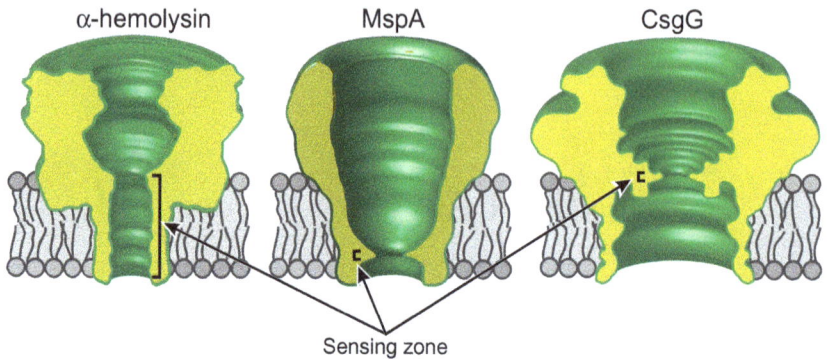

Figure 4.1. Cross-sections of three channel forming proteins. The structural contour of each protein has been determined by crystallographic methods. The channel diameter in each of these nanopores varies throughout its length, but in one region each narrows to form a ~0.75 nm radius aperture. As explained in the text, this narrow region of the nanopore is the channel's primary sensing zone. Recombinant DNA methods have been used to molecularly engineer MspA and CsgG to allow passage of ssDNA and to optimize their electrical signals for sequencing.

Structure and Assembly

As shown in Figure 4.1, the common architectural feature of these nanopores in a lipid membrane is a single nano-dimensioned channel lined primarily with hydrophilic amino acids. Even though the cross-sectional area of the channels varies throughout the conduit's length, at least a portion of each channel narrows down to a radius of ~0.75 nm. As explained below, this narrow region of the channel is a critical feature that makes such proteins excellent sensors of nucleic acid molecules.

A second common feature of protein nanopores is that each is composed of several polypeptide chains — the monomer subunits — which self-assemble and insert into lipid-bilayer membranes. There are seven identical monomers in α-hemolysin, eight in MspA, and nine in CsgG. That the monomer polypeptide chains can assemble into a continuous aqueous channel through an otherwise hydrophobic membrane makes it possible for any laboratory to easily create a nanopore sequencing setup. Self-assembly also greatly facilitates manufacture of inexpensive devices containing hundreds or thousands of identical nanopores that can function in parallel.

The commonality of these three proteins also extends to the structure of their secondary, tertiary, and quaternary features. As with all proteins, the same hydrophobic effects that stabilize self-assembly of amphiphiles (see Chapter 3) play a central role in how their tertiary and quaternary features are assembled. Because the importance of these hydrophobic effects have best been characterized for α-hemolysin — where assembly can be studied *in vitro* outside of the *Staphylococcus aureus* cells in which the protein's subunits are synthesized — it serves as a model for similar hydrophobic effect-dependent processes that usually occur intracellularly for other proteins such as MspA or CsgG that assemble in the cell's own membrane after synthesis.

The α-hemolysin monomers are synthesized by *S. aureus* bacteria and secreted as a water-soluble toxin that readily inserts into the lipid bilayers of cell membranes or synthetic lipid bilayers. The monomers perform no known function within *S. aureus* cells but instead have evolved to inhibit the growth of other cells that compete for nutrients. Even though the secreted monomer is highly soluble in aqueous solutions and its primary structure is polar with no obvious surface regions that are predominantly hydrophobic, seven of these monomers on a cell membrane can assemble into a heptameric nanopore with a stem region that penetrates through the membrane's hydrophobic bilayer. The stem region firmly anchors the nanopore to the membrane through which it creates an aqueous channel. Because the assembled nanopore is not easily released from the membrane, assembly of the water-soluble monomer to form the water-insoluble nanopore must involve important conformational changes. These conformational changes are driven largely by hydrophobic effects which are schematically outlined in Figure 4.2.

As initially synthesized in the aqueous cytoplasm of a *S. aureus* cell, the antiparallel β strands that make up the bulk of α-hemolysin are folded in such a way that the component hydrophobic amino acids are packed into the interior of the molecule. Thus folded, the aqueous solvent-accessible surface of the monomer contains mostly polar and

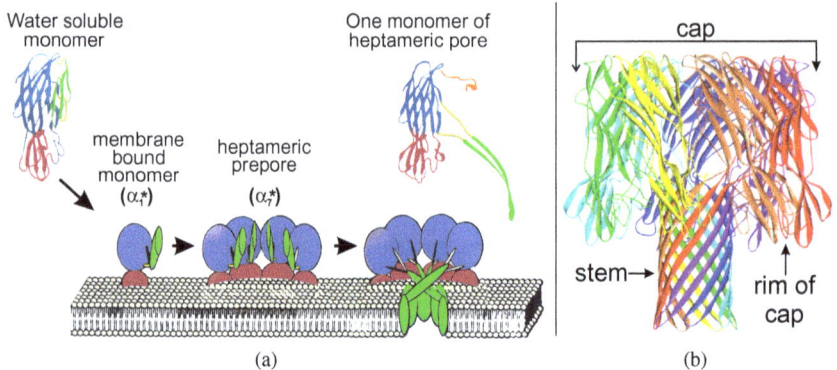

Figure 4.2. α-hemolysin assembly. (a) A ribbon diagram (*top left*) of the monomer shows how the two predominantly hydrophobic β-strands (green) that will contribute to the nanopore stem are folded against other β-strands that will form the nanopore's cap (blue). The β-strands that will contribute to the cap's rim (red) interact with a bilayer by associating with the lipid head groups to form the membrane-bound monomer shown in the assembly schematic (*bottom, schematic, left*). After a ring of seven monomers forms a heptameric prepore (schematic, middle), the 14 β-strands that will be the nanopore's stem unfold from the monomer's cap and cooperatively form the boundary of the nanopore's stem through the lipid bilayer. The ribbon diagram (top, right) shows how the two β-strands (green) from each monomer unfold to contribute to the nanopore's stem. Identical colors characterize the regions of each monomer in the ribbon diagrams and in the schematic (redrawn from Ref. 1). (b) Ribbon diagram of an assembled nanopore with each of its seven polypeptide chains in a different color (redrawn from Ref. 2).

charged amino acids whereas the hydrophobic amino acids that are packed against each other in the interior of the protein are not solvent accessible. This initial folding pattern maintains the aqueous solubility of the monomer and also prevents premature assembly of the heptameric nanopores until after secretion by *S. aureus* and interaction with hydrophobic chains of lipid bilayers.

The secreted monomers bind to cellular membranes, but can also bind to synthetic lipid bilayer membranes. The membrane associated monomers are mobile and diffuse within the plane of the membrane and drift randomly together with the fluid lipids (see Figure 3.6). As more of these mobile monomers associate with the membrane and diffuse around on the membrane surface, they naturally collide into each other. Eventually, seven of these monomers collide to form a ring-shaped circle of monomers that can assemble to form a hepta-meric prepore.[3] Because assembly of seven monomers into a hepta-meric prepore occurs in less than 5 ms (lower order ring-shaped inter-mediates with fewer than seven monomers have not been detected), the rapid process that produces the prepore heptamer remains poorly understood. But once the prepore heptamer is produced, 14 β strands (two from each monomer) unfold and extend through the bilayer to associate with each other, forming the completely assembled, mem-brane-bound nanopore.[1,3]

Sensing Polynucleotides and Identifying Nucleobases

The fundamental function of the nanopore used in sequencing is to rapidly transduce several physical and chemical properties of individual nucleobases into easily measured ionic currents. The measured cur-rents are those flowing through the nanopore's aqueous channel while nucleic acid polymers are driven through the channel, altering its elec-trical resistance in a nucleobase-specific manner.

As explained in Chapter 2, the total electrical resistance, R, of a nanopore's channel when it contains only an ionic solution should be determined to a large extent by the solution's resistivity, ρ, the chan-nel's length, L, and the channel's cross-sectional area, A, that is filled by an ionic solution, i.e. $R_{total} = \frac{\rho}{A}\left(L + \frac{\pi}{2}r\right)$. Because the cross-sectional area of the channel varies as a function of the radius *squared*, the elec-trical resistance of the channel when filled with an ionic solution will be dominated by the resistance through the length of its narrowest radius region. The small ~0.75 nm radius of this narrow region forces the DNA to traverse as an extended string, just as a cotton thread can be pulled through the eye of a small needle whereas folded thread or tangles will be extended or untangled before traversing. As a DNA strand is driven through the nanopore, each nucleobase of the DNA strand obstructs the nanopore to a different, characteristic degree. The amount of current which can pass through the nanopore at any given moment therefore varies depending on whether the nanopore is blocked by an adenine (A), thymine (T), guanine (G), or cytosine (C). In practice it is not one nucleotide that blocks the nanopore but a section of the DNA that includes several nucleobases. The group of nucleotides is called a "*k*mer", where *k* approximates the number of nucleobases that simultaneously contribute to the recorded signal from the narrow region of the nanopore. The blocked current through the nanopore indicates that a strand of DNA is passing through, and small changes in the current level represents a direct reading of the *k*mer of nucleobases that occupy the ~0.75 nm radius region of the pore.

Early experiments in which different length polynucleotides were driven through a nanopore's channel provided important clues about which region of the channel is most sensitive to the presence of the traversing polymer and about the contour of the DNA during its traversal through the nanopore. For example, when polynucleotides containing 12 or more nucleobases are electrophoresed through α-hemolysin by a 120 mV bias, the current magnitude during a blockade is reduced to just ~8–10% of the open-nanopore magnitude when no polynucleotides are in the pore (Figure 4.3). Although shorter

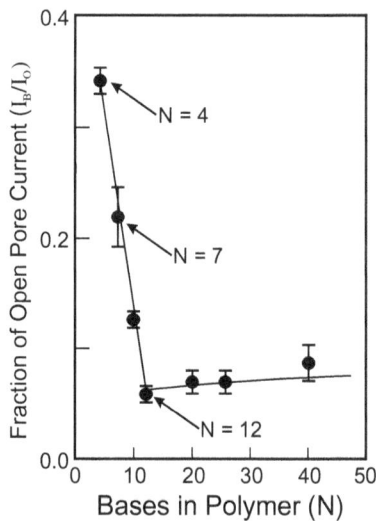

Figure 4.3.   Fraction of open pore current during blockades in α-hemolysin *vs.* number of bases in a traversing polymer. Adapted from Ref. 4.

polymers produce blockades of lower amplitude (see Figure 4.3), longer polynucleotides that are composed of more than 12 monomers reduce the current to about the same ~8–10% value of the open-pore current magnitude observed with a 12-monomer polynucleotide.

The above observations inform us about the channel region that is most sensitive to the presence of the traversing polymer and about the contour of a polynucleotide as it is driven through the nanopore. As explained in Chapter 1, long polynucleotide strands are flexible polymers that usually form disordered, roughly spherical tangles in ionic solutions. But to traverse through the narrowest confines of a nanopore channel whose radius (~0.75 nm) barely exceeds the radius of a single extended polynucleotide strand (~0.6 nm), a portion of the polynucleotide at least the length of this narrow radius region must be fairly straight. The length of an extended polynucleotide can easily be calculated by multiplying the number of nucleotides in the polymer by the extended distance from one nucleobase to the next nucleobase. This distance is ~0.45 nm, which means that a polynucleotide composed of 12 monomers would be ≈5.4 nm long (12 × ~0.45 ≈5.4 nm) to cause the maximal current reduction to ~8–10% of the open α-hemolysin nanopore. Since far longer polymers do not cause greater current reductions than the 12 monomer long polynucleotides, the sensing region in α-hemolysin must not extend beyond ~5.4 nm. And because polynucleotides shorter than 12 nucleobases have less of an effect on blocking ionic current, it follows that all of the nucleotides in the ~5.4 nm length of a 12 nucleobase polymer must contribute to the increased resistance of the nanopore. Stated differently, if we interpret the pore's resistance as a measure of the fraction of the pore volume from which ions are excluded by the translocating polymers, it follows that for polymers shorter than the pore's sensing length, the resistance values will vary with the polymer length, while for polymers that are longer than the pore's sensing length, they will be polymer length independent.

Since ~5.4 nm is remarkably close to the crystallographically determined 5.2 nm length of α-hemolysin's narrow stem region, these observations suggest that this entire stem region of the nanopore is the zone that senses the passage of a polynucleotide. The total length of the α-hemolysin channel through the stem and the cap is 10 nm, so polynucleotides that are longer than 10 nm would extend through both the stem and the nanopore's 4.8 nm length cap above the stem. This means that the cap region does not significantly contribute to polynucleotide sensing.

Other experiments using tethered polymers that travel measured distances into the MspA and CsgG nanopores confirm that in these nanopores, it is also primarily the narrow ≈0.75 nm radius region of the pore that is sensitive to the presence of the traversing DNA. But these nanopore regions have much shorter lengths than the α-hemolysin stem, so that to a first approximation the ~4–5 nucleobases centered in the nanopore channel's narrow region dominate the increased resistance that reduces the nanopore's ionic current during a blockade. These observations, together with some simple calculations of the percentage of the channel's cross-sectional area occupied by a polynucleotide traversing through any of the three nanopores depicted in Figure 4.1, provides further evidence for believing that the narrow ~0.75 nm radius region of the nanopore is the zone primarily responsible for sensing and identifying the sequence of nucleobases in the polynucleotide. Such calculations show that a 0.6 nm diameter polynucleotide would occupy ~65% of the area in the narrow ~0.75 nm radius region of a nanopore's channel. While the channel entrance into α-hemolysin's cap region narrows to a radius of 1.4 nm, an entering polynucleotide would occupy only 18% of this area. In MspA and CsgG nanopores, there are no regions other than the narrow ~0.75 nm zone of the channel in which a traversing polynucleotide would occupy even 10% of the channel's cross-sectional area. Thus, if area displacement of the conducting ionic solution is to account for the decreased current during a blockade, only the narrow ~0.75 radius zone in these 3 nanopores can account for a major fraction of the current reductions observed during a blockade.

Since both the cross-sectional area and the length of the narrowest cross-sectional region are important in determining the total resistance of an ionic solution-filled channel, it is to be expected that the current reductions observed during translocation of long polynucleotides through the much shorter narrow regions of MspA or CsgG would be less than in α-hemolysin. This is in fact what is observed: instead of blockades that reduce $\geq 90\%$ of the open pore current through α-hemolysin, maximal blockades reduce ~80% and ~75% of the open pore currents in MspA a CsgG, respectively.

While occupancy of the nanopore by the volume of a polynucleotide may account for some of the current reduction observed during a blockade, resistance calculations based on ionic solution displacement show that volume occupancy alone can explain only a fraction of the decreased ionic current observed during polynucleotide translocation. More importantly, the observed current resistances when homopolymers of DNA's 4 different nucleobases traverse through the nanopore cannot be correlated with the area or volume differences among the nucleobases. This is because ionic behavior and electrical resistance within the very tightly confining space of a polymer-filled nanopore channel is actually far more complex than can be explained by volume occupancy and the simple resistance calculation of Eq. (2.9). Multiple properties and phenomena whose interactions remain poorly understood influence the nanopore's resistance to ionic flow. These include hydration effects, polymer conformation, electrostatic effects, nanopore perimeter charge effects, and the possibility of differing interactions of the different nucleobases with the amino acids that line the channel's perimeter.[5] But despite the fact that many of these complex interactions remain poorly understood in the confining space of a nanopore, experimental evidence demonstrates that a nanopore does in fact distinguish among the nucleic acid bases and their sequences and transduces their identity into easily measured current levels. By driving polynucleotides of known sequence through a nanopore, these current levels have been correlated with their corresponding nucleobases so that the sequence of nucleobases in an unknown polymer can be determined.

References

1. Kawate, T. & Gouaux, E. Arresting and releasing Staphylococcal α-hemolysin at intermediate stages of pore formation by engineered disulfide bonds. *Protein Sci.* **12**, 998–1006 (2003), doi:10.1110/ps.0231203.
2. Song, L. *et al.* Structure of Staphylococcal α-hemolysin, a heptameric transmembrane pore. *Science* **274**, 1859–1866 (1996).

3. Thompson, J.R., Cronin, B., Bayley, H. & Wallace, M.I. Rapid Assembly of a Multimeric Membrane Protein Pore. *Biophys. J.* **101**, 2679–2683 (2011).
4. Meller, A., Nivon, L. & Branton, D. Voltage-driven DNA translocations through a nanopore. *Phys. Rev. Lett.* **86**, 3435–3438 (2001).
5. Zwolak, M. & Di Ventra, M. Physical approaches to DNA sequencing and detection. *Rev. Mod. Phys.* **80**, 141–165 (2008).

Chapter 5

Helicases and DNA Motor Proteins

Alicia K. Byrd and Kevin D. Raney

A critical step that made nanopore sequencing possible was the development of methods to drive single strands of genomic DNA through a nanopore in single steps, each the length of a deoxynucleotide monomer. Earlier experiments had shown that when the voltage bias required to drive a polynucleotide through a nanopore is applied, the average time one nucleobase takes to translocate through the pore is less than 10 μs. This very rapid transport rate made it impossible to resolve and identify the four nucleobases because the small picoampere current differences between A, T, C, and G were masked by the nanopore's inevitable electrical noise. As discussed in Chapter 2, the precision required to measure the small currents through a nanopore imposes a fundamental physical limitation that can only be overcome by slowing the rate of DNA translocation, i.e. increasing the length of time each nucleobase remains in the nanopore.

Calculations showed that an adequate signal-to-noise ratio would require each nucleobase to reside in the nanopore's sensing region for more than 100 μs. Several attempts to slow DNA's transport rate by physical means such as decreasing the voltage bias through the membrane or increasing the solution viscosity failed because they did not sufficiently lengthen the time each base resided in the sensing region and often introduced problems of their own. For example, increasing medium viscosity can slow polynucleotide transport through the nanopore but also reduces the rate of ion flow through the nanopore, so the signal-to-noise ratio is not significantly improved. In fact the issue of how best to move a polynucleotide through the nanopore is further complicated because it is now recognized that nucleobase identification can be optimized by advancing the polynucleotide in discrete steps.

The Role of Motor Enzymes in Nanopore Sequencing

Eventually, it was realized that processive motor enzymes such as polymerases or helicases could not only slow the movement of a polynucleotide strand through a nanopore but also ratchet the motion so that each base paused in the pore for a few milliseconds. Several molecular motor enzymes move along the length of a polynucleotide strand in steps that correspond to the distance from one nucleobase to the next. Movement involves work and motor enzymes perform this work by harnessing the free energy released during catalytic hydrolysis of high energy phosphate bonds in nucleoside triphosphates (NTPs) or deoxynucleoside triphosphates (dNTPs). Depending on which of several proteins is chosen and the conditions that are selected, motor proteins step along DNA strands at rates from 10 to 1,000 bases/s. This range includes those that are slow enough to enable reading of each nucleobase but fast enough for sequencing to proceed rapidly. The motor enzyme must also generate sufficient force to translocate a DNA or RNA strand against a competing electrophoretic force of the applied voltage. Furthermore, its association with the strand must be stable enough for the polynucleotide to undergo multiple catalytic cycles such that very long DNA or RNA molecules can be translocated through the nanopore. Finally, the diameter of the motor enzymes must be greater than the diameter of the nanopore so that it does not get pulled into the channel.

Polymerase Motors

If a motor enzyme satisfies all of these conditions, it will bind to a single-stranded polynucleotide that can then be captured and rapidly pulled into the nanopore by the applied voltage. Translocation will be retarded or completely stopped once the bound motor moves

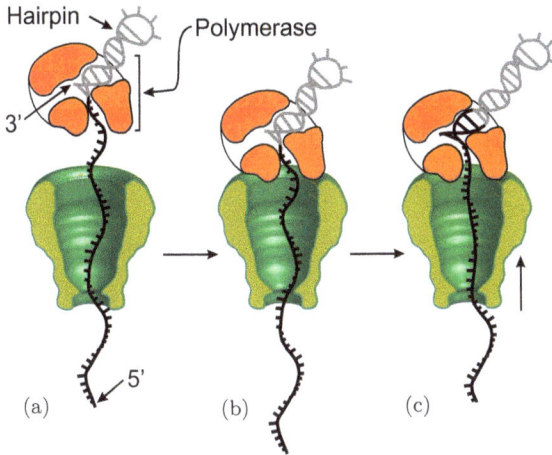

Figure 5.1. Using a polymerase to step DNA through a MspA nanopore. Short hairpins (grey) are ligated to the 3′ ends of the polymers that are to be sequnced (black) to provide the primer-template site at which the DNA polymerase will bind and initiate polymerization.). (a) The DNA with a bound polymerase is captured and drawn into the nanopore. (b) The polymerase motor is activated as it makes contact with the nanopore through which it cannot pass. (c) The now activated polymerase adds new nucleobases to the 3′ hydroxyl terminus of the primer as it steps 3′ to 5′ along the template strand that is gradually pulled back out of the nanopore into which it had initially been driven as shown in (a).

into contact with the nanopore's aperture through which it cannot fit. Thereafter, a motor enzyme can implement slow, stepwise translocation of the polynucleotide through the nanopore. Depending on which of several motor enzymes is used and its characteristic direction of stepping, the polynucleotide can be slowly pulled back out of the pore as the motor enzyme steps through the length of the polynucleo tide (Figure 5.1).

The first successful demonstration showing that a nanopore can actually read the sequence of nucleobases in a polynucleotide used a DNA polymerase as the motor enzyme.[1] The polymerase pulled the DNA strand out of the pore against the electrophoretic force of the voltage bias used to initially capture and pull the strand into the nanopore (Figure 5.1). DNA polymerases are motor enzymes that catalyze the assembly of precursor dNTPs to form a new strand of DNA that base pairs with a pre-existing template strand of DNA. Of several polymerases that were tested, *phi29* polymerase proved able to remain tightly associated with the DNA against the load imposed by the voltage bias that initially had pulled the DNA into the pore (Figure 5.1(a)).[2]

Polymerases bind and initiate polymerization of a complement strand only at a primer–template junction. To provide the primer site at which the DNA polymerase will bind, a short hairpin is ligated onto the 3′ terminus of the template polymer whose sequence is to be determined. The DNA with a bound polymerase is then placed in the *cis* chamber for sequencing. Clever hairpin design inhibits the polymerase activity in the *cis* chamber until after one of these template–hairpin complexes with a bound polymerase is drawn into the nanopore (Figure 5.1(a)) and the polymerase enzyme comes into contact with the nanopore whose small channel diameter prevents its entry (Figure 5.1(b)). Once in contact with the nanopore, the polymerase is activated and begins to incorporate (and hydrolyze) the appropriate dNTPs that base-pair with the template strand. As the polymerase continues moving through the template strand and synthesizes the complementary strand, the entire length of the template DNA is stepped out of the nanopore (Figure 5.1(c)). To pull the template strand that is being sequenced through the nanopore sensor against the force of the electric field acting in the opposite direction, the enzyme harnesses the energy released as each dNTP is hydrolyzed and incorporated into the growing complementary strand.

Helicase Motors

The MinION uses helicases as a convenient method of controlling the movement of DNA and RNA through nanopores. Helicases are motor proteins responsible for separating double-stranded DNA (dsDNA) into single-stranded DNA (ssDNA) (Figure 5.2). *In vivo*, each strand is then acted upon by a number of enzymes to synthesize a based-paired strand, leading to two dsDNA daughter strands. *In vitro* during MinION sequencing, the helicase also separates any dsDNA

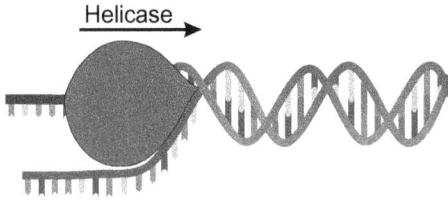

Figure 5.2. Helicase function. Helicases separate duplex DNA or RNA into two single strands.

in the unknown sample into ssDNA and then delivers the strand to which the helicase is bound to the nanopore at a single-nucleotide stepping rate that makes it feasible to identify the sequence of nucleobases in the single-stranded polynucleotide.

In contrast to the polymerases that have been employed to pull ssDNA through nanopores against the force of the electric field acting in the opposite direction, helicases are used to limit the rate at which the electric field drives a strands of DNA through the nanopore (Figure 5.3). Because most helicases bind to ssDNA, a short negatively charged synthetic polymer with a bound helicase is ligated onto the terminal nucleotide of one strand of the dsDNA whose sequence is to be determined. This negatively charged polymer with a bound helicase serves as the starter strand that is electrophoretically translocated from the *cis* chamber into and through the nanopore together with the dsDNA to which it is ligated (Figure 5.3(a)). This brings the helicase into contact with the nanopore where it resides during sequencing (Figure 5.3(b)). Contact with the nanopore activates the helicase which, in the presence of ATP in the *cis* chamber, then initiates movement along one strand of the dsDNA, separating the dsDNA into a single strand which is stepped through the nanopore as it is unzipped from its complementary strand (Figure 5.3(c)). Because the helicase diameter is larger than the nanopore channel, further translocation of the starter strand and attached ssDNA through the nanopore proceeds at the stepping rate of the helicase as it continues to separate the dsDNA into one strand that translocates through the nanopore and a complementary strand that is left in the *cis* chamber. The helicase

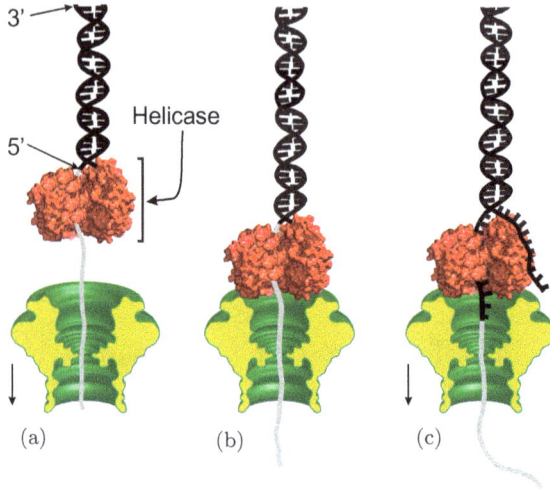

Figure 5.3.   Using a helicase to step DNA through a CsgG nanopore. (a) The end of a synthetic anionic polymer (grey) that is ligated onto one of the strands in much longer dsDNA (black, but whose full length is not shown here) is captured and pulled into the nanopore. (b) The helicase is activated as it makes contact with the nanopore through which it cannot pass. (c) The now activated helicase steps 5′ to 3′ along one strand of the dsDNA that will translocate through the nanopore as the complementary strand is displaced into the surrounding solution.

steps the ssDNA through the nanopore at a rate that makes it feasible for the nanopore to read its nucleobase sequence at rates approaching 500 bases per second.

Biological Roles of Helicases

Because helicase motors play such an essential role in nanopore sequencing, it is worth knowing more about their structure and function in living cells. Moreover, some of these *in vivo* activities of helicases may in the future find applicability to nanopore sequencing. DNA replication is only one of the many *in vivo* processes that require helicases. Other processes that involve nucleic acid processing by helicases include DNA transcription into RNA, translation of RNA into proteins, DNA repair, and RNA processing. Helicases are called molecular motors because they couple binding and hydrolysis of ATP to perform the work of moving on a polynucleotide while 'unzipping' duplex RNA or DNA. Key to this chemical transduction activity is the directional movement of a helicase along one of the polynucleotide strands in a 5′–3′ manner or in a 3′–5′ manner. Each specific helicase will move in only one direction. The directionality can help define the specific role played by a helicase. The unidirectional step-like movements of a helicase are a critical feature of nanopore sequencing, helping to control the DNA strand's translocation through the nanopore in a stepwise manner.

To perform their many functions *in vivo*, helicases interact with numerous other proteins. Furthermore, DNA in cells does not exist as a naked double helix, but instead is bound to proteins. In eukaryotic cells, DNA is packaged into chromatin by wrapping around proteins called histones. These complexes, known as nucleosomes, are separated by short DNA linkers. DNA compaction into nucleosomes is beneficial for storage both because it allows the long DNA polymer to fit within the space of the nucleus and because it serves to protect the DNA. Mechanisms to allow access to the DNA are also necessary, and one function of a family of enzymes closely related to helicases is to serve as motors that reposition or remodel nucleosomes on DNA. These chromatin remodelers do not perform the canonical helicase function of separating strands of duplex DNA but instead move along dsDNA in an ATP-dependent manner and remove the histone from the DNA. Multiple cycles of ATP hydrolysis and translocation result in the repositioning of the histones on the DNA, thus allowing access to different regions of the chromosome to other proteins and enzymes during transcription, replication or repair.

There are also non-histone DNA-bound proteins in cells such as transcription complexes. These proteins must be removed before unwinding of DNA can occur during DNA replication. Bound proteins which are not removed may cause stalled replication forks which can result in breakage of the DNA strands or incomplete genomic duplication. For this reason, bound proteins must be able to be removed from the DNA in a process that requires energy both to separate the DNA strands and to disrupt the interaction between the protein and the DNA. These bound proteins are often displaced by the replicative helicase. But when the nucleoprotein complex resists removal, accessory helicases may function to displace the protein barriers. RNA polymerases that catalyze transcription are the most frequently encountered bound proteins, but helicases are involved in removal of all types of proteins bound to DNA including the complexes that stabilize telomeric DNA.

Helicases also manipulate RNA and participate in all aspects of RNA metabolism. All phases of transcription of DNA into RNA, including elongation, termination, and splicing require RNA helicases. Although RNA is generally single stranded, RNAs fold into complicated three-dimensional structures that include the formation of short regions of duplex RNA. RNA helicases can also serve as "RNA chaperones" by unwinding duplex regions that form incorrectly, thus allowing the RNA to re-fold into the appropriate final structure. The ribosome is a large molecular machine composed of both protein and RNA subunits. RNA helicases are involved in both ribosome biogenesis and translation of RNA into proteins by the ribosome. Additionally, transport of RNA within the cell, degradation of RNA, and dissociation of proteins from RNA are also all dependent on RNA helicases.

Helicase Structure and Action Mechanisms

Helicases can be divided into six superfamilies based on amino acid sequence motifs in the primary sequence. Superfamily 1 and 2 helicases function as monomers or dimers, while those in superfamilies 3–6 form hexameric rings. To date, only the superfamily 1 and 2 helicases, such as Dda from bacteriophage T4 (Figure 5.4) have been employed in nanopore sequencing, and its structure and action mechanisms will be emphasized here.

Superfamily 1 and 2 helicases are composed of a monomer with two domains often referred to as domains 1A and 2A (Figure 5.5), or helicase domains 1 and 2, and any number of accessory domains that vary between helicase families.[4] The ssDNA interacts with many regions of domains 1 and 2 and the ATP binding site occurs at the interface of these domains. Important to the mechanism for DNA

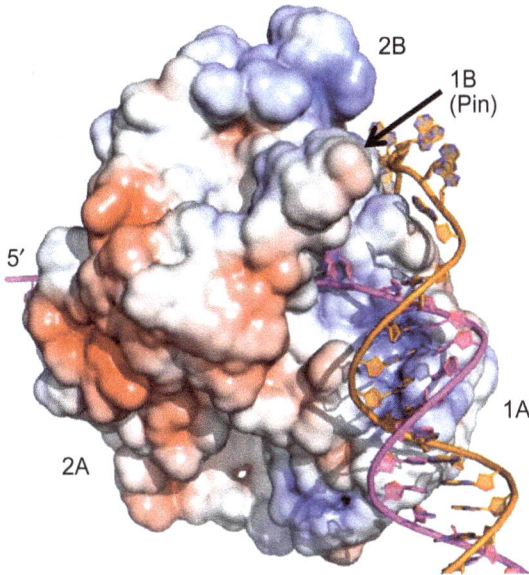

Figure 5.4. Model of Dda with duplex DNA. Space filling representation of Dda with negatively charged amino acids in red, positively charged amino acids in blue, and neutral amino acids in white. The DNA strand on which Dda translocates is shown in purple and the displaced strand is orange. The purple strand to which the helicase binds and on which it moves passes through a channel formed between domains 1A and 2A and the 1B/2B domains (top). Note the 5′ end of this strand exiting at the left. Image originally prepared by Steve White and adapted from Refs. 3 and 4. Used with permission.

unwinding is the "pin" domain shown in red in Figure 5.5.[4] The two strands of dsDNA split at the pin domain as one is moved within the Dda molecule where it binds along the surfaces of domains 1A and 2A while the displaced strand passes along the outer surface of the helicase as shown in the surface rendering in Figure 5.4.

As mentioned above, most helicases bind to ssDNA, or to a portion of dsDNA that is single stranded, and then unwind the dsDNA by translocating toward the duplex. Helicases contain amino acid sequence motifs which vary slightly from one helicase family to another, but serve similar purposes (Figure 5.6). Because the helicase called Dda has been successfully engineered to function in nanopore

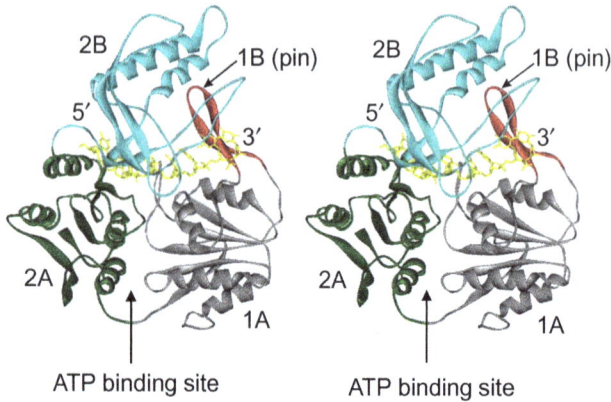

Figure 5.5. Stereo of Dda helicase. Crystal structure of Dda helicase (PDB code: 3UPU; Ref. 4) bound to single-stranded DNA (yellow). The two helicase domains, 1A and 2A, are shown in gray and green, respectively. ATP binds in the cleft between domains 1A and 2A. A region of domain 1A forms the 1B or pin domain (red) and an insertion in domain 2A forms the 2B domain (cyan).

sequencing, we will focus here on this monomeric motor enzyme with a detailed description of its interaction with and translocation along DNA.

The key motifs in Dda are involved in binding to DNA and ATP, and coupling the energy of ATP hydrolysis to movement along the DNA strand (Figure 5.6).[5] Helicases move in discrete steps and in the simplest case, the energy from hydrolysis of a single ATP molecule moves the helicase by a single nucleotide along the DNA. In some cases, hydrolysis of one ATP molecule may result in translocation by more than one nucleotide, or multiple rounds of ATP hydrolysis may be required for the helicase to move by one nucleotide on the DNA. The number of nucleotides translocated by the helicase in a single step is referred to as the physical step size. For many helicases, including Dda, this has been measured to be a distance equivalent to one nucleotide.

The manner in which Dda helicase moves on a strand of DNA as it unwinds dsDNA is referred as an "inchworm" mechanism (Figure 5.7). Unidirectional translocation of the helicase along the DNA is due to conformational changes associated with ATP binding and hydrolysis that sequentially drive movement by one nucleotide

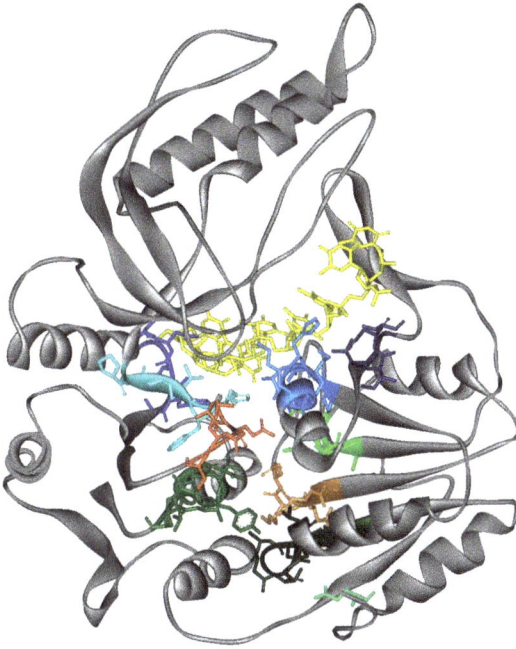

Figure 5.6. Key helicase motifs. The structure of Dda is shown with the amino acid motifs which bind DNA colored in blue, motifs which bind ATP colored in green, and motifs involved in coupling ATP hydrolysis to translocation on DNA colored in orange.

along the DNA. The helicase monomer contains two domains that comprise the DNA binding sites. Each of these domains contains part of the DNA binding site, and domain closure upon binding of ATP causes movement of the rear domain towards the front domain in the direction of translocation (Figure 5.7). Release of ADP and P_i after ATP hydrolysis causes the cleft to open, resulting in movement of the forward domain, thereby completing a cycle of ATP hydrolysis-powered translocation. If the helicase is moving on dsDNA, this directional movement by one nucleotide results in separation of one base pair of the duplex.

Wild-type Dda is not a highly processive enzyme. Processivity is the number of nucleotides unwound (or translocated) before dissociation of the helicase. This property is essential for DNA sequencing

Figure 5.7. Dda moves like an inchworm on a DNA strand. The backbone of a DNA strand drawn as a rainbow colored bar to aid seeing movement of the Dda schematic (blue). The inchworm mechanism in Dda involves movement of its 2A domain in response to ATP binding and movement of 1A domain upon release of ADP and P_i.

because a higher processivity means a greater number of nucleobases can be moved through the nanopore in a single continuous read. If the helicase falls off a long piece of dsDNA after advancing only partway through its length, the remainder of the strand would not be read by the nanopore. To assure that complete strands of DNA can be read in a single continuous run, the helicase is engineered for use in the MinION to greatly enhance Dda's processivity. Consequently, continuous reads of megabase lengths of DNA have been achieved.

Molecular Mechanism for Coupling ATP Hydrolysis to Movement

Crystallographic structural analysis of Dda and closely related helicases in multiple conformations and in the presence and absence of bound nucleotides provide insight into Dda's translocation mechanism. In the absence of ATP, eight nucleotides of DNA are bound across the 1A and 2A domains. Interactions between Dda and the DNA are predominantly hydrogen bonds and ionic interactions of the protein with the backbone. These interactions are weak, allowing them to form transiently, thus enabling the rearrangements necessary for ATP hydrolysis and DNA translocation by inchworm-like movement to occur. Two amino acids in the 1B or pin domain, phenylalanine 98 (F98) and proline 89 (P89), stack with the sixth nucleotide (T6) in the binding site (Figures 5.8(a) and 5.8(b)) through interactions of electrons in orbitals of the amino acids with those in the nucleobase. A different stacking interaction has been observed in the closely related RecD2 helicase which shows the fifth base rotated into a pocket in

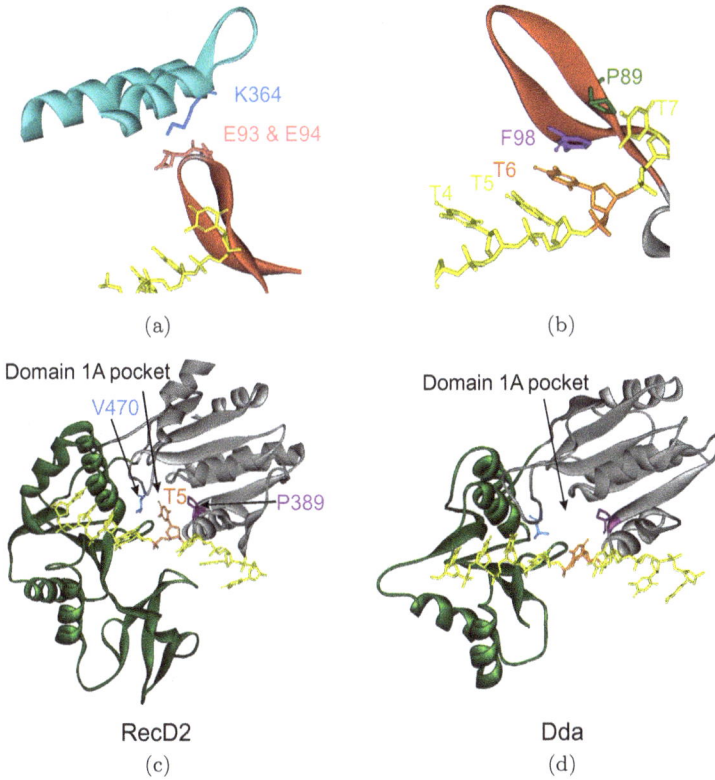

(a)

(b)

(c)

(d)

RecD2

Dda

Figure 5.8. The DNA binding site. (a) The crystal structure of Dda (PDB code: 3UPU; Ref. 4) reveals an interaction between two glutamic acid residues (E93 and E94, pink) at the tip of the pin (red) with a lysine residue (K364; blue) in the 2B domain (cyan). This interaction is critical for coupling ATP hydrolysis to DNA unwinding. (b) Two amino acids on the pin (phenalanine 98, purple and proline 89, green) stack with the 6th base (orange) bound by Dda. This interaction is also important for coupling ATP hydrolysis to DNA unwinding. The structures of a closely related helicase, RecD2 (c) and Dda (d) show different conformations of the DNA binding site that likely exist at different moments in the catalytic cycle in both proteins. In the RecD2 structure, the 5th base (orange) is flipped up into a pocket in domain 1A (gray) formed by amino acids in helicase motif 1A (proline 389, purple) and motif III (valine 470, blue) whereas in the Dda structure, a similar pocket is formed by proline 62 (purple) and valine 152 (blue), but the base is not present in the pocket.

domain 1A (Figure 5.8(c)). In the case of Dda, this base is not rotated into the pocket (Figure 5.8(d)). Nevertheless, it is probable that this interaction occurs as the catalytic cycle progresses in both RecD2 and Dda but the crystal structures probably captured the DNA at different stages of the cycle. Both the stacking of a base with amino acids in the pin domain and rotation of a base into the pocket of domain 1A allow the bases to serve as cogs that prevent sliding of the protein along the DNA.

Binding of ATP in the cleft between domains 1A and 2A causes a closure of these domains, which results in movement of the 2A and 2B domains one nucleotide towards the 1A and 1B domains (Figure 5.9). The 1A domain remains tightly bound to the DNA and does not move in response to ATP binding. Before ATP binding, one base is flipped out relative to the remainder of the bases and rests in the pocket formed by helicase motif 1A. During the conformational change due to ATP binding, the base in the motif 1A pocket flips to motif III. This is followed by ATP hydrolysis and opening of the domain 1A/2A cleft. Upon ATP hydrolysis, domains 1A and 1B move forward by a single nucleotide. Thus, one round of ATP binding, hydrolysis, and release results in translocation by one nucleotide along the DNA and concomitant separation of one base pair. The movements of Dda relative to the dsDNA that the helicase is unzipping can be seen in the movie at https://www.worldscientific.com/worldscibooks/10.1142/10995#t=suppl.

The stepping motion of the helicase is important to nanopore sequencing because the current measurements corresponding to each base are highly sensitive to this motion as bases pass through the pore. Therefore, it is important to fully characterize the stepping motion. For example, there are actually two half-steps within a single-nucleotide step: domain closure in response to ATP binding, and domain opening upon release of ADP and P_i. For some enzymes, such as Hel308, each of these half-steps is visible when analyzed using nanopores, resulting in an apparent step size of 0.5 nucleotides per step.[6] This results in twice as many changes in current levels for the same DNA sequence relative to enzymes which translocate the DNA in single nucleotide steps. The duration of alternating levels is independent of ATP concentration while the time spent at intervening

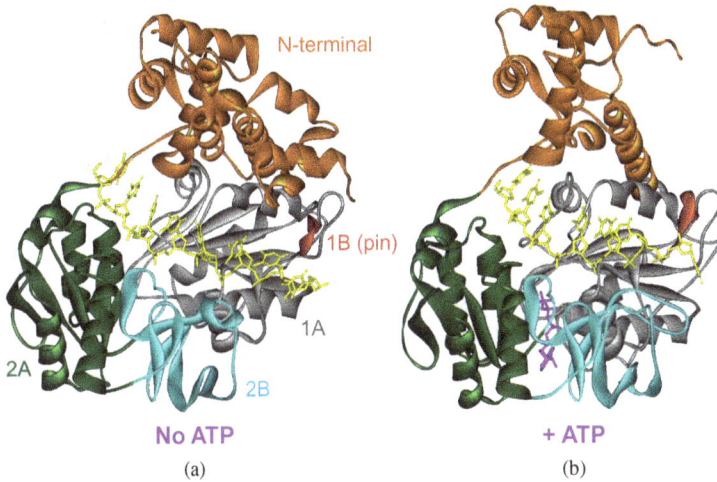

Figure 5.9. Domain closure upon binding of ATP. (a) Because the crystal structure of Dda with and without ATP has not been documented, the crystal structure
of a protein with similar motifs and functions — RecD2 (PDB code: 3GP8) — is
shown here bound to single-stranded DNA (yellow). Domains are colored similarly
to Dda in Figure 5.5, with domains 1A and 2A in gray and green, respectively. The
pin (domain 1B) is red and domain 2B is cyan. The N-terminal domain of RecD2
(orange) is absent in Dda. (b) Structure of RecD2 bound to DNA and an ATP analog
(purple) illustrates the conformational change in the 1A and 2A domains to form the
ATP binding site.

levels is dependent on the ATP concentration. Hence, the helicase
conformational change due to ATP binding and that due to ATP
hydrolysis appear as discrete current levels during nanopore sequencing. One of the observed steps is likely due to the actual movement
of the enzyme by one nucleotide along the substrate. The other may
be due to repositioning of the DNA/helicase complex relative to the
pore as a result of conformational changes within the helicase.

References

1. Manrao, E.A. *et al.* Reading DNA at single-nucleotide resolution with
a mutant MspA nanopore and phi29 DNA polymerase. *Nat. Biotechnol.*
30, 349–353, (2012), doi:10.1038/nbt.2171.

2. Cherf, G.M. *et al.* Automated forward and reverse ratcheting of DNA in a nanopore at 5-Å precision. *Nat. Biotechnol.* **30**, 344–348 (2012), doi:10.1038/nbt.2147.

3. Aarattuthodiyil, C., Byrd, A.K. & Raney, K.D. Simultaneous binding to the tracking strand, displaced strand and the duplex of a DNA fork enhances unwinding by Dda helicase. *Nucleic Acids Res.* **42**, 11707–11720 (2014).

4. He, X. *et al.* The T4 phage SF1B helicase Dda is structurally optimized to perform DNA strand separation. *Structure* **20**, 1189–1200 (2012), doi:10.1016/j.str.2012.04.013.

5. Byrd, A.K. *et al.* Dda helicase tightly couples translocation on single-stranded DNA to unwinding of duplex DNA: Dda is an optimally active helicase. *J. Mol. Biol.* **420**, 141–154 (2012), doi:10.1016/j.jmb.2012.04.007.

6. Derrington, I.M. *et al.* Subangstrom single-molecule measurements of motor proteins using a nanopore. *Nat. Biotechnol.* **33**, 1073–1075 (2015), doi:10.1038/nbt.3357.

Chapter 6

Development of Multipore Sequencing Instruments

James Clarke

Previous chapters of this book have discussed the important features and processes that explain why a nanopore in an ion-impermeable membrane can determine the sequence of nucleobases in one or more strands of DNA or RNA. As will be detailed in subsequent chapters, when the genome of an entire organism is to be sequenced, the organism's dsDNA must first be extracted and prepared for analysis.

During this extraction process, the DNA of interest usually undergoes some shearing that brings down the fragment length from the naturally occurring lengths of double-stranded DNA (dsDNA) in the organisms chromosomes. Typical extractions can result in DNA sheared to 30–100 kilobase-pairs (kbp) length fragments, although with careful handling (see Chapter 7) much longer fragment lengths are possible.

The extracted dsDNA fragments are then adapted through a library preparation process that makes them suitable to be sequenced by the MinION. The adapted DNA sample is introduced into the MinION as a solution containing the mixture of fragments. An array of up to 512 nanopores begins to capture one end of each fragment which is drawn into the nanopore by an applied voltage. As the dsDNA fragments are unwound by a helicase enzyme, the nanopore generates a current signal modulated by the nucleobase sequence (see Figure 1.5). The current signals are then decoded to a sequence using a base-calling algorithm. Once read and decoded, the sequences can be assembled to represent the original continuous DNA sequence of the organism.

A process in which multiple different analytes (here, the DNA fragments) are simultaneously characterized in a single run is called a multiplex assay, and the MinION instrument is an example. In a MinION instrument, all of the operational nanopores capture analyte DNA strands at random from the common pool. The sequences of these strands are determined in parallel for each nanopore. Multiplex assays are commonly used when a large number of analyte molecules are to be characterized in a reasonable length of time. This is of course the usual situation whenever the entire genome of an organism is to be sequenced. Even a small genome, such as that of *E. coli*'s single circular chromosome, contains ~4.6 Mbp (~4,600,000 base pairs) of DNA. If the genomic DNA from a population of these bacteria is sheared into 45 kbp, there would be at least ~100 different fragments of the genome whose sequence would need to be read.

As with any measurement, errors can occur in the course of reading the sequence in a strand of DNA, and there are a number of possible sources of error in nanopore sequencing. For example, the action of the motor enzyme may move bases through the nanopore too rapidly to be measured, or the current signals of different sequences may be so similar that the nanopores cannot distinguish between them. It is also possible that the DNA may be damaged or modified in a way that has not been trained into the base caller.

While the single molecule accuracy of the platform has improved rapidly over the last few years, there is often a need to average over some of the single molecule errors to produce a higher overall accuracy. An increased consensus accuracy is produced when multiple single molecule reads of the same region of the genome are compared. The number of repeat reads that are "piled up" for the consensus is called the depth. For human sequencing, it is common to use depths of 30× to 60× to generate a consensus. For instance, to sequence the 4.6 megabase *E. coli* genome to a depth of 30×, 138 megabases of sequence information are required. Assuming that a single nanopore can read the DNA at 450 bases per second (b/s), and that the nanopore is utilized for sequencing 90% of the time, it would take ~3.5 days to acquire 138 megabases of sequence. This calculation illustrates why a single nanopore channel is not sufficient for many applications and why a multiplexed sensor array was always considered key to producing a commercially viable nanopore sequencer.

The first multipore recordings were achieved with a four-well device, each well covered with a lipid bilayer containing an α-hemolysin nanopore (Figure 6.1). The device contained a Ag/AgCl (silver/silver chloride) electrode in a shared *cis* chamber that delivered its contents to all four nanopores. Each of the four wells, which served as four *trans* chambers, contained an individually addressable Ag/AgCl electrode connected to one of four sensitive patch clamp amplifiers. Although this was a simple device, it proved that nanopores could be arrayed, that there was no observable cross-talk between the four wells, and that each of the four nanopores acted independently of each other.[1] There has been much refinement to the array design, but the core concept of a single shared *cis* chamber and multiple separate *trans* chambers has remained the same since those early experiments.

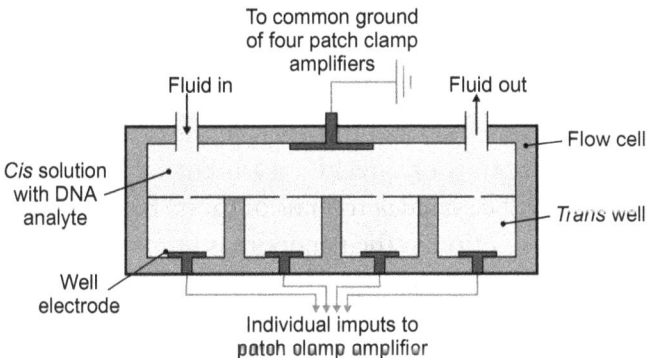

Figure 6.1. Initial four-well device for testing purposes to show that arrays of membranes with nanopores could be formed and that they operated independently of each other.

A key requirement in the product development was to replace the bulky and expensive patch clamp amplifiers and associated equipment with compact, less expensive Application-Specific Integrated Circuits (ASIC). The ASIC is built into the MinION's flow cell which contains a shared *cis* chamber and an array of 2,048 *trans* chamber wells. The ASIC used in the MinION's flow cell has the ability to record the ionic current through 512 pores at any one time.

Continuing progress in membrane stability made it possible to provide the MinION with a "factory-formed" bilayer array. The lipid bilayer membrane was replaced with a more stable silicon-based triblock co-polymer (see Chapter 3 and Figure 3.5), the array design was changed to add features for controlling the fluids, and the Ag/AgCl electrodes were replaced with platinum electrodes for stability and ease of manufacture. As explained below in the description of the electrochemistry, the platinum electrodes contact a ferri/ferrocyanide mediator solution in their immediate vicinity to provide an electrochemical reaction with a very fast turnover while ion-selective films prevent this electrochemical mediator solution from mixing with the larger solvent volumes that carry the polynucleotide analytes in the *cis* and *trans* chambers.

The first MinION made available to the public was the MAP (MinION access programmer) device consisting of a base unit for connecting to a personal computer via a USB cable, and a disposable flow cell. The MinION Mk1b used the same flow cell, but the base unit was redesigned. The original separate lid of the device was now hinged and a clip was introduced to ensure good thermal contact between a heat sink and the MinION's fan and the ASIC in the flow cell.

Sequencing depends on precise measurement of the nanopore's ionic conductivity during analyte translocation. Because the conductivity of the ionic solution in the nanopore is temperature dependent, uncontrolled temperature changes would impact correct interpretation of the nanopore currents. The temperature of the flow cell is therefore controlled by dissipating the heat that its ASIC generates throughout the MinION's metal casing. Another key feature was improved magnetic shielding which reduced high-frequency electrical noise from the fan. This was particularly important because changes were made to the sequencing chemistry that improved the helicase speed from 70 to 450 b/s.

Flow Cells and Sequencing Chemistry

Although the MinION can be used to detect a broad range of analytes, the first applications were focused on strand sequencing of either DNA or RNA molecules. The process of transforming extracted DNA for a sequencing run is called library preparation. The first library preparation kit added adapters to the sample DNA and was ligation based. The adapters provided a long "leader" section to thread the polynucleotide into the nanopore, a site to allow tethering of the sample to the membrane for improved sensitivity, an engineered helicase designed to ratchet the nucleotides through the nanopore one base at a time, and chemistry that assures helicase activation only upon nanopore contact. Detailed explanations about the various preparation kits and diagrams showing how they modify the DNA in preparation for sequencing are described in Chapter 8.

The library preparation kits have undergone constant improvement, including genetic and chemical modification to the helicase motor for more controlled DNA movement. It was discovered that the helicase motor protein could run much faster, which was exploited in parallel with improvements to the base-calling algorithm. Mutations were added to the nanopore protein where it contacted the helicase motor, which made the signals more consistent and by 2017 raised the base-call accuracy above 90% even at 450 b/s. In 2016, a transposase-based kit was made available which simultaneously fragmented the sample DNA and attached the adapters in one step (SQK-RAD001, SQK-RAD002). This had the added advantage of reducing the time needed to run the kit, bringing library preparation times down to just 10 min.

In 2016, the structure of R9, the current MinION nanopore, was revealed. R9 is a mutant of the CsgG lipoprotein from *E. coli* which *in vivo* transports polypeptides across the bacterial membrane. The CsgG lipoprotein is composed of nine identical subunits that assemble into the beta-barrel pore described in Chapter 4. The pore was engineered to allow DNA, instead of peptides, to be translocated through the structure. The CsgG pore increased base-calling accuracy with further gains coming from a new version of the base-calling algorithm. This was based on a recurrent neural network (RNN) which can learn features of the data to exploit. The RNN better represents the underlying physical process of a DNA strand interacting with the nanopore, giving a single molecule 1D accuracy over 92%. (The 1D term refers to a single continuous read of one DNA strand.) As the R9 series developed over time, the helicase was modified to increase its maximum speed to ~450 b/s. The adapter was redesigned to reduce long-lasting pore blockages, and beads were added to the running solution to further reduce blocking and improve throughput. These and other developments raised the maximum sequencing output expected from each flow from less than 1 GB in 2014 to around 20 GB in 2017.

With the commercial release of the MinION in 2015, base calling was at first provided as a cloud-based service, but there was an obvious user preference for local base calling to enable real-time in-field applications. The high single molecule accuracy of 1D combined with its 10 minute Rapid kits library preparation made this possible, and in 2017 local base calling was enabled as part of the MinKNOW software for real-time analysis. The advent of $1D^2$ chemistry (see Chapter 8) in 2017 increased base call accuracy to >97% with throughput speeds of ~450 b/s.

Technical Description

The MinION instrument is an electronic device that provides the interface between the user's PC and the nanopore sensor in a synthetic amphiphilic layer (Figure 6.2). It also provides power to the ASIC that controls the nanopore's sensing function, controls the temperature, shields the nanopore sensor from electronic noise, and transfers data to the PC. The device is around the same size as a small candy bar, measuring 105 mm × 33 mm × 23 mm. Power and data transfer is achieved through a single USB 3.0 port which provides the power required by MinION and the ASIC (5 V @ 900 mA max). Data transfer rates on USB 3.0's are sufficient to handle all 512 recording channels that send raw data at 33 kHz.

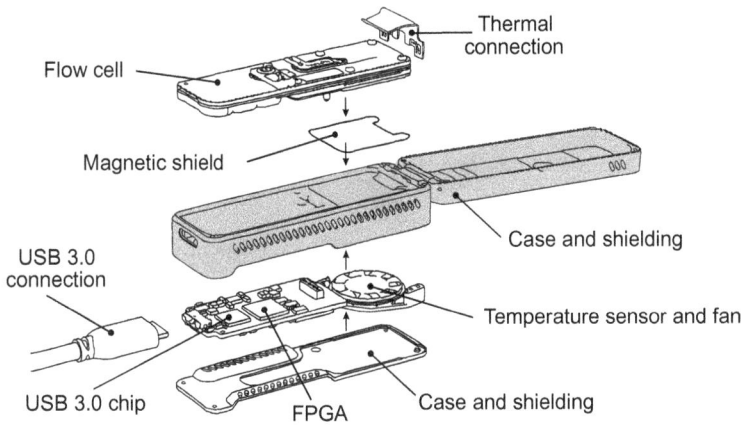

Figure 6.2. Exploded view of the MinION device and flow cell.

Although the MinION does not have a dedicated heating element, heat from the electrical energy input to its ASIC is dissipated throughout the MinION casing. The MinION's electronics include a temperature sensor and associated control program to maintain the flow cell and its contents at a constant temperature during a sequencing run. Excess heat generated by the ASIC is dissipated to the ambient air by a small fan. Heat from the ASIC and control of the fan speed are sufficient to allow the MinION to maintain the flow cell at a steady temperature of 34°C.

The metal case of the MinION is electrically grounded to the flow cell's printed circuit board (PCB) and protects the nanopore sensors from electric fields. Continuity between the flow cell's PCB, MinION case enclosure, and lid is provided by magnetic contacts and a conductive hinge assembly. Magnetic shielding is built into the MinION case to further reduce noise in the signal. A field programmable gate array (FPGA) in the MinION is used for data handling and control.

Description of ASIC

The ASIC is a high-density array of low-noise amplifier circuits that is used to measure and provide a digital read out of the current flow

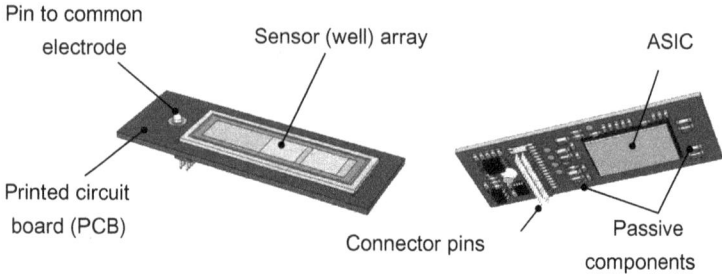

Figure 6.3. Top (left) and bottom (right) view of the chip used in the MinION flow cell showing the sensor array, printed circuit board (PCB), and application-specific integrated circuit (ASIC).

between each of the *trans* compartment electrodes and the electrode in the common *cis* chamber (Figure 6.3). In addition, the ASIC receives commands from the host software system to provide control functionality for the sensor array including acquisition frequency, signal filtering, sensor current range, multiplex input selection, electrode bias potential generation, and deselection of sensor inputs with broken membranes to avoid saturation of the measurement circuits. The ASIC is bonded to the lower side of the flow cell's PCB substrate with connections through to the upper side where the sensor array is mounted to ensure a short connection path with minimal parasitic capacitance. Other passive components to support the ASIC are present on the PCB, while features like temperature control are handled by the MinION device and MinKNOW software that operates in the attached computer.

The MinION flow cell ASIC is designed to operate across a wide range of operating conditions depending on the application. Typical operations have an acquisition frequency of 2–20 kHz while measuring 10–1,000 s of picoampere currents. The applied electrode bias potential can be controlled using scripts in MinKNOW with a range of ± 1 V.

The MinION ASIC also has the ability to use the applied potential to unblock any of the channels that are obstructed with tangled DNA or contaminants. A reversal of the potential can "flick" any stray DNA or contaminants out of the pore on a per channel basis and reset the channel to an "open pore" state to allow the next strand to be sequenced. This ability is further utilized in "Read Until" schemes, where the MinION can be programmed to respond to partially sequenced fragments, choosing to either complete sequencing all captured DNA fragments or reject some fragments back into the common *cis* chamber to save time for higher priority fragments during a run. This feature, unique to nanopore sequencing, enables the user to analyze only DNA strands that contain predetermined signatures of interest. The sequence of the first few bases in a DNA strand passing through the nanopore is analyzed in real time. If found to be of interest, sequencing continues. If found to be of no interest for the particular experiment at hand, the strand can be rejected from the pore, freeing that pore to sequence preferred strands.

Description of the Sensor Array

The sensor area of the flow cell is comprised of 2,048 individual sensor wells, each designed to hold a single nanopore. These sensor wells are a micro patterned structure sitting on a silicon substrate and created by photolithographic techniques (Figure 6.4). The patterned struc-

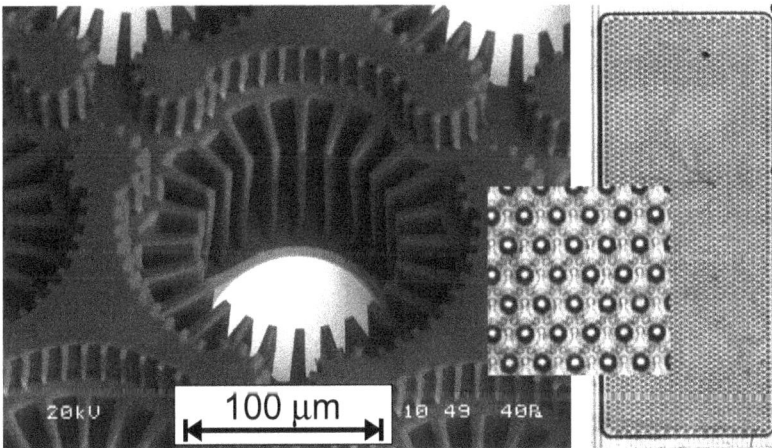

Figure 6.4. Scanning electron microscope image of a single sensor well (left) and the hexagonal pattern of wells in the array (right).

Figure 6.5. Multiplex layout of the nanopore sensor array for the MinION flow cell.

ture of each well facilitates stable membrane formation and insertion of a single nanopore per well. The membrane provides an insulating layer between the *trans* chamber wells and the overlying shared *cis* chamber so that once a nanopore is inserted, communication between the electrodes is dominated by the signal through that pore. The MinION uses a proprietary amphiphilic polymer and a synthetic pre-treatment oil to form the membrane (see Chapter 3 and Figure 3.5). Each well is approximately 90 μm deep with a platinum electrode at the base of the well. This electrode is then connected electrically to the PCB through the silicon layer.

The MinION flow cell currently has 2048 active well electrodes organized hexagonally on the surface of the array. The active wells are arranged in two blocks of 32 × 32 with a set of four inactive wells separating them (Figure 6.5). There are also inactive wells present at the edges of the array blocks that are covered with a gasket. Although the ASIC is capable of recording through only 512 ASIC channels simultaneously, there are 2,048 active well electrodes organized with multiplexing circuitry into groups of four. To maximize the yield of channels containing a single pore and improve the sequencing output of the consumable, the choice of a single nanopore from each group of four is controlled by the multiplexing circuitry that selects the best 512 wells as the primary group whose output will be selected (a process called "mux selection"). During a sequencing run, as some of the well membranes break down or nanopores become irreversibly clogged, other groups of wells can be selected.

MinION Flow Cell

The MinION flow cell is a disposable element of the sequencing platform that provides the fluidic interface between the sensor and the electrodes, allowing samples to be analyzed. The flow cell is composed of a molded plastic fluidic chamber, fluidic gasket, the sensor array, a common reservoir, common electrode, ion-selective membrane, a PCB and an application-specific integrated circuit (Figure 6.6). The sensor array is made up of many sensing wells, each designed to record through a single nanopore as described earlier.

The flow cell itself is designed with a number of distinct areas. These areas include: fluid inlet, value, inlet track, sensor array, exit track, and waste track. The latest versions of the flow cell also have a SpotON port added so that small volumes of sample can be added directly on to the array. The volume of the *cis* chamber area over the sensor array is 100 μL, while the waste can hold 2 mL. Using the SpotON inlet, sample volumes of 75 μL can be used either by a dropwise application, or by pipetting straight above the sensor array.

The flow cell chemistry is preassembled before being shipped. The membranes and nanopores are made and inserted into the membranes in a production facility. When the flow cell is received, the user is asked to perform a "platform qc" or "Check my flow cell" to ensure that the flow cell meets the minimum requirement of nanopores. The script takes approximately 5 min and examines each of the active wells for

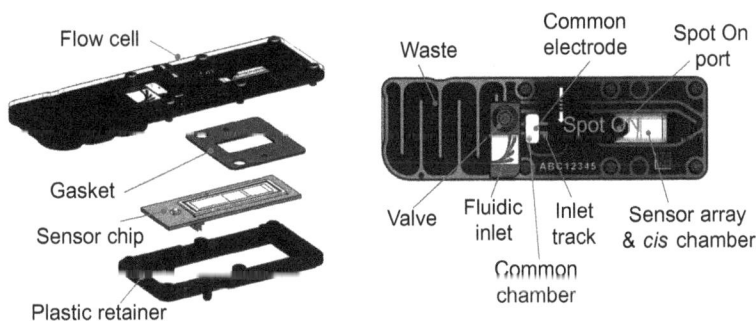

Figure 6.6. Exploded view of the MinION flow cell (left) and a top view with the functional areas labeled (right).

the integrity of the membranes, the presence of a nanopore, and the ability of the nanopore to detect a simple test analyte. As received, the system contains ferri/ferrocyanide mediator solution in the shared *cis* chamber and well chambers. The *cis* chamber also contains a dilute solution of a DNA test analyte composed of a single-stranded DNA (ssDNA) attached to a quadruplex DNA structure. When the ssDNA translocates through the pore, the quadruplex unfolds under applied potential. The process of unfolding the quadruplex structure slows the DNA's translocation through the pore so that the translocation process can be measured more easily. The appearance of the expected signature blockades is a quality control check that the flow cell is working correctly.

Once the flow cell has been checked, the user is asked to perfuse the *cis* chamber of the test analyte with run buffer. The run buffer contains a simple salt solution, magnesium chloride, and adenosine triphosphate (ATP) that serves as the substrate for the helicase motor. The flow cell is then ready for the user to add their adapted library and start the sequencing run.

Description of Electrochemistry

The nanopore signal is essentially a measure of ionic current through the pore that is modulated by nucleobases when a single-stranded nucleic acid is translocated through the pore. The electrochemistry is chosen to be a two-electrode self-referencing system for which a minimum of two electrodes are needed, one on each side of the membrane. One of these electrodes is common to all wells (positioned in the center of the flow cell), while the second electrode is in the base of each active well. The common electrode is isolated in the common chamber of the flow cell by an ion-selective film that prevents the electrochemical mediator solution from leaking into the *cis* chamber.

The academic researchers who developed the first nanopore techniques used silver/silver chloride electrodes because of their fast electron transfer and stable reference potential. The MinION flow cell employs a different electrode system comprising platinum electrodes and an electrochemical mediator. This allows the sensor chip to be fabricated with standard inert metals, while the mediator provides a redox couple that facilitates rapid electron transport at the platinum surface for a highly responsive measurement. The electrochemical mediator currently used is a bright yellow potassium ferricyanide/ferrocyanide solution. For sequencing, a negative potential of -180 mV is applied to the common electrode while the well electrodes of the sensor array are held at a virtual ground by the amplifier input. The half-cell equations for this couple during a sequencing run are as follows:

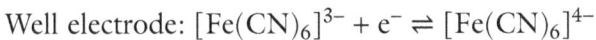

$$\text{Well electrode: } [\text{Fe}(\text{CN})_6]^{3-} + e^- \rightleftharpoons [\text{Fe}(\text{CN})_6]^{4-}$$

$$\text{Common electrode: } [\text{Fe}(\text{CN})_6]^{4-} \rightleftharpoons [\text{Fe}(\text{CN})_6]^{3-} + e^-$$

As an ionic current passes through a nanopore between electrodes, the concentrations of the redox pair are perturbed, causing a gradual change in the potential difference according to the Nernst equation. Although this voltage change is partially corrected in the sequencing scripts, the capacity of the mediator couple is ultimately depleted and the flow cell stops working. The flow cell is then returned to the company for refurbishing, but it can produce many gigabases of sequence data during its lifetime.

Recent Updates and Future Developments: Direct RNA Sequencing

RNA sequencing is central to understanding some of the most important cell functions related to processing of genetic information. The usual approach is to convert the RNA into cDNA, which can then be sequenced to determine the RNA sequence. Users can perform cDNA sequencing on the MinION with single-molecule long reads, high-throughput, and low numbers of PCR cycles. A significant advance is that MinIONs are also capable of reading an RNA strand directly without copying it into cDNA. This allows the researcher to interrogate modified bases, or to accurately count transcripts without reverse transcript or PCR biases. Details of direct RNA sequencing will be described in Chapter 8, and so this will only be briefly summarized here. To read an RNA strand, just a few modifications need to be made to the nanopore system, mostly in the helicase motor and adapter chemistry. An essential discovery was a motor enzyme that is processive on native RNA strands with step-wise movement. This enzyme works in the opposite direction to the DNA helicase, and therefore the RNA strand is threaded through the nanopore 3′–5′, making the electrical signals obtained from each nucleobase quite different from those seen with DNA that translocates 5′–3′ through the nanopore. Despite these changes, a neural network base-calling algorithm has been developed and an accompanying publication[2] describes the method by which the same nanopores and flow cell used for DNA can sequence RNA with the MinION. A special kit designed for direct RNA nanopore sequencing is available from Oxford Nano-pore Technology. Typical single molecule accuracies of >90% were obtained with long read lengths dependent on the quality of the sample. A number of academic groups have recently published research results involving direct RNA sequencing, showing that different types of modified bases could be observed directly.[3]

Advanced Nanopore Sequencing Platforms

The MinION sequencing system is designed so that each nanopore channel is operated independently. This means the platform can be arranged in a number of different formats to suit a range of applications. The other products that use the same sequencing chemistry as the MinION include the GridION, PromethION, SmidgION, and Flongle. The PromethION is designed for larger projects, such as sequencing the human genome at high coverage. Runs of over 90 Gbases of data have been obtained from a single PromethION flow cell. A PromethION instrument contains docking for 48 flow cells and is capable of recording through 144,000 nanopores in parallel. The PromethION flow cells use the next generation of ASIC, with 12,000 wells and 3,000 simultaneous recording channels. Currently, PromethION is in early access phase with sequencing runs being performed at a number of institutions.

The throughput of the MinION has improved since its release, with runs of 15 Gbases of sequences from a single flow cell. While some researchers might require more data, there are also a large number of applications where this is in fact more information than is required. In response to a growing demand for a less expensive product having fewer channels, a flow cell dongle (or "Flongle") plugs into the MinION and allows use of a smaller inexpensive disposable flow cell containing 128 channels. At current output, this would generate 1–2 Gbases of data per flow cell, which would be sufficient for applications such as pathogen identification, antimicrobial resistance profiling, metagenomics analysis, and targeted panels.

The SmidgION takes the smaller flow cell and builds a device around it. With a custom ASIC and fewer channels, the device is small enough to fit onto a mobile phone that can access a cloud-based informatics pipeline. Given that DNA extraction and library preparation can be performed in 10 min with a kit, and base-calling algorithms available online, in the near future it will be possible to address complex genomic questions using nothing more than a nanopore-based sequencer such as the SmidgION attached to a cell phone.

References

1. Reid, S.W., Reid, T.A., Clarke, J.A., White, S.P. & Sanghera, G.S. Formation of layers of amphiphilic molecules. WO Patent 2009077734 (2007).
2. Garalde, D.R. *et al.* Highly parallel direct RNA sequencing on an array of nanopores. *Nat. Meth.* (2018), doi:10.1038/nmeth.4577.
3. Smith, A.M., Jain, M., Mulroney, L., Garalde, D.R. & Akeson, M. Reading canonical and modified nucleotides in 16S ribosomal RNA using nanopore direct RNA sequencing (2017). Available on: https://www.biorxiv.org/content/early/2017/04/29/132274, doi:10.1101/132274.

Chapter 7

DNA Extraction Strategies for Nanopore Sequencing

Joshua Quick and Nicholas J. Loman

Introduction

Read lengths in nanopore sequencing depend largely on library preparation rather than any limitation of the sequencing chemistry. Long
reads are useful for many applications, but in particular *de novo* assembly because long reads are able to span repeats in the genome, thereby
increasing the usable length of assembled contiguous sequences.[1]
The longest reads generated by nanopore sequencing now exceed 1
megabase pairs (1.2 Mbp) at time of publishing, but even longer reads
will likely be achievable with further improvements in DNA extraction and library preparation methods. They will be extremely helpful
in assembling difficult regions of the genome such as eukaryotic centromeres and telomeres. It may be possible one day to sequence entire
bacterial chromosomes or even eukaryotic chromosomes in a single
read! Possibly the only limit to read length is the rate of naturally
occurring single-strand breaks in DNA.

This chapter will describe the most useful extraction techniques
for nanopore sequencing and focus on best practices for routine work,
experimental design, and quality control, which are summarized in
Figure 7.1. Finally, we will discuss ongoing efforts to generate "ultra-
long reads".

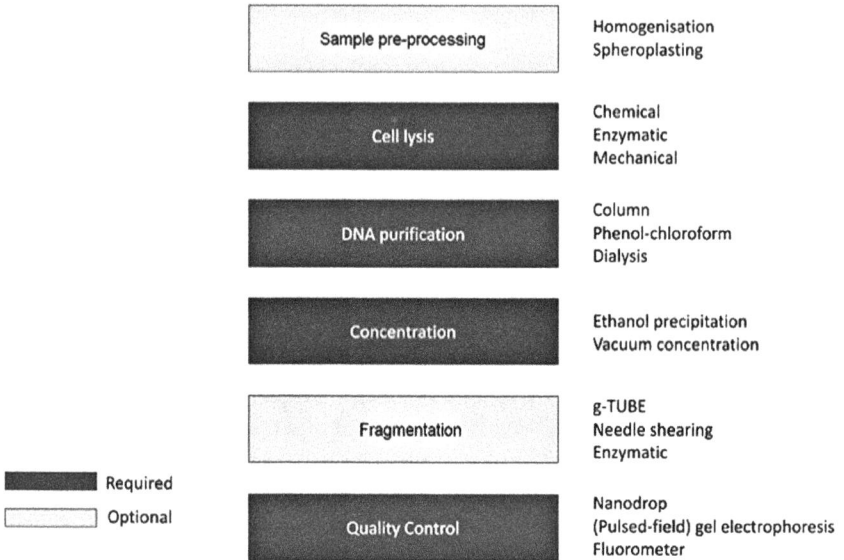

Figure 7.1.   Steps for DNA extraction. Required and optional sample pre-processing and fragmentation steps are shown.

Choosing a DNA Extraction Strategy

Multiples strategies are available for extracting DNA from various sources and each produces different lengths (Figure 7.2). While it may be tempting to pick strategies that optimise isolation and purification of high molecular weight DNA, this comes at a significant cost in terms of time and labor. Sample input, read length, and cost are all highly interdependent factors and designing a good experiment first requires an understanding of how these factors affect the outcome. If the goal is to assemble a bacterial genome (for example, to produce a reference sequence), obtaining reads above the "golden threshold" of 7 kb (the length of the ribosomal RNA operon) will in most cases produce a finished circular genome with no gaps. The importance of the ribosomal RNA operon is that it is typically the longest repetitive region in a bacterial genome, so having reads longer than this threshold will enable these repeats to be "anchored" to unique parts of the genome, permitting their assembly. Therefore, for many bacterial genomes, a simple spin column extraction (yielding typically up to 60 kb fragments) would be appropriate because these fragment sizes are sufficient to generate the read required read length.

But if the collective genomes of a mix of closely related species or strains from an environmental sample — a metagenome — is being sequenced (an extremely challenging assembly problem), then longer reads are important for strain reconstruction during assembly. Similarly, complex genomes such as the human genome will benefit from the longest possible reads due to extensive repeating sequences such as those in centromeres, some of which remain largely unassembled years after the announcement of the first human reference genome. In these cases, there is plenty of source material in the form of cells and tissues, and so it is reasonable to attempt a high molecular weight DNA extraction.

Other applications may be limited by input quantity. Many clinical and environmental samples have intrinsically low biomass. In order to extract sufficient DNA for sequencing, these must be extracted with efficient high yield recovery methods such as magnetic beads or spin

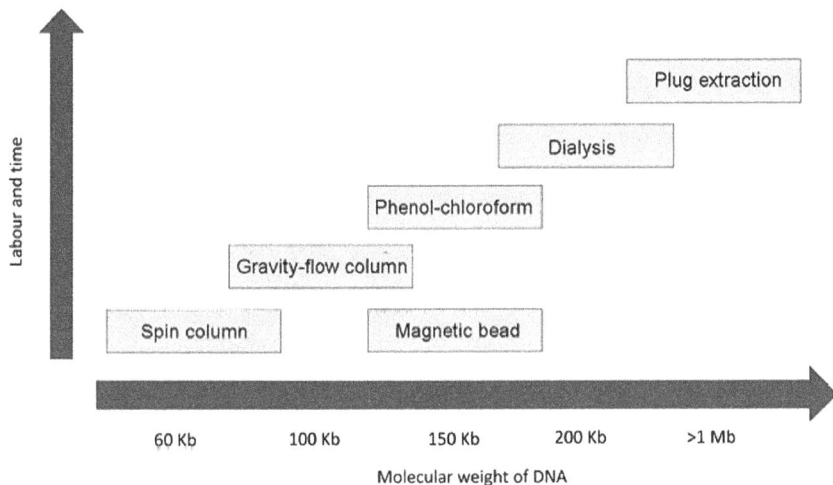

Figure 7.2. Average size of DNA fragments isolated by different methods discussed in this chapter.

columns in which shearing can easily occur. An understanding of the biological question to be addressed and the amount of sample available are therefore key to designing a good sequencing experiment.

DNA Extraction Kits

Hundreds of DNA extraction methods have been described in the literature. Often they have been developed for specific sample types, but they share common steps such as cell lysis, purification, and elution/precipitation. Here we will describe some of the routine methods used in DNA extraction.

The simplest way to get started is to use any of several commercial DNA extraction kits that offer a high level of consistency and excel for small amounts of samples. They are more expensive than manual methods, typically costing around $5 per sample, and fragment length will be limited to around 60 kb. Spin column kits are the most common type of DNA extraction kit and use either silica or anion exchange resins to reversibly bind DNA so that it can be separated from cellular proteins and polysaccharides that do not readily bind to the column. Samples are added to the top of a small tube, then forced through a binding matrix in the tube during centrifugation. In some cases, the columns include lysis reagents so that release of the DNA from cells, binding to the matrix, followed by washing and eluting the DNA can be done in about an hour. Many extractions can be done in parallel by using multiple positions in the centrifuge rotor.

It is worth understanding how spin columns work, but also their weaknesses. Most kits use high concentrations of guanidinium hydrochloride in the lysis buffer. Guanidinium is a chaotropic agent that disrupts the hydrophobic interactions between water and other molecules. It lyses cells, denatures membrane proteins, and precipitates DNA by disrupting the hydration shell which maintains DNA's solubility in aqueous conditions. Under these conditions, DNA adheres to the binding matrix in the column while proteins and other contaminants pass through. The DNA bound to the silica resin membrane can then be washed with 70% ethanol to remove any remaining proteins and salts, as well as the lysis buffer itself. After washing, DNA is released from the matrix by adding a low ionic concentration buffer such as 10 mM Tris and incubating for a few minutes. The DNA is soluble in the aqueous solution and can then be eluted from the column by centrifugation.

For common Gram-negative bacteria (such as *E. coli*) > 60 kb can be extracted using a kit with spin-column extraction in less than 30 minutes. Spin columns have a binding capacity of about 5–10 μg and can be run in batches, making them suitable for extracting large numbers of samples. An important point is that genomic DNA is sheared by the physical forces generated when it is centrifuged and forced through the porous resin.

Shear forces are reduced in gravity flow columns such as the Genomic-tip (Qiagen). These employ the same binding technology as spin columns but come in larger sizes, the largest of which has a binding capacity of 500 μg. Gravity flow columns are not centrifuged but instead left upright in a rack which allows the lysate/wash solutions to drip though by gravity. These kits recover DNA with an average size of 100–200 kb but are much slower. Unlike spin columns, DNA is eluted from the column in a large volume, then precipitated with isopropanol to concentrate it. Gravity flow kits are especially useful for isolating large quantities of DNA which is likely to be higher quality than that produced with spin columns, so this method may be an appropriate choice for certain nanopore applications.

Magnetic beads with a wide variety of functional groups on the surface have many uses in molecular biology. Beads used for isolating genomic DNA are polystyrene and magnetite microspheres with a coating that incorporates carboxylate groups. In the presence of a chaotropic agent, DNA in solution transitions into a condensed 'ball-like' state which is attracted to the beads. This allows the DNA to be purified by washing the beads with ethanol in a low ionic strength solution, and then releasing the DNA by a dilute buffer which causes the negative charges of the carboxylate groups to repel the similarly charged DNA off the beads. The main advantage of magnetic beads is speed of processing because DNA binding occurs very quickly in solution and the beads are rapidly harvested by attraction to a magnet. Such techniques are also amenable to automated handling and are used in many commercial high-throughput robot platforms.

Manual Techniques

Certain sample types, particularly plant and animal tissue, cannot be directly lysed and must be broken up by a process called homogenization. This is usually done by freezing with liquid nitrogen, then grinding in a cold Dounce homogenizer or mortar and pestle. The liquid nitrogen has a dual purpose of making the sample very brittle for efficient grinding but also inhibits native nuclease activity which would degrade DNA.

For samples such as bacteria and yeast, spheroplasting is used to digest the cell wall while keeping the cell intact in sucrose buffer to protect them from osmotic shock. The name *spheroplast* derives from the spherical appearance of cells after cell wall digestion. This process allows cells with thick walls such as yeast and plant cells to be easily lysed by the addition of a detergent that disrupts the cell membrane.

Cell Lysis

Cell lysis is the process of breaking open cells to release DNA. This is usually performed with detergents, enzymes, or physical methods. Bacteria, yeast, plants, and animals have very different cellular structures, and therefore different lysis methods are employed. Commonly used detergents include sodium dodecyl sulfate (SDS) for bacterial and mammalian cells, and cetyltrimethylammonium bromide (CTAB) for plants. Strong detergents like SDS also serve to protect DNA from degradation by inactivating nucleases. Many Gram-positive bacteria are too tough to lyse with detergents due to their peptidoglycan cell wall, and so lysis solutions may also incorporate additional enzymes such as lysozyme that breaks down the cell wall by hydrolyzing specific chemical bonds in peptidoglycan. Other enzymes are used for *Staphylococcus* (lysostaphin) and *Streptomyces* (mutanolysin) where lysozyme is ineffective. Yeast cell walls are composed of two layers of β-glucan which requires lyticase and zymolase to break down.

Some spore-forming bacteria and fungi may also have additional layers of peptidoglycan or chitin, making them extremely resistant to enzymatic or chemical lysis so mechanical methods may be required.

The most common method uses various sizes of 'beads' made from glass or zirconium which are vigorously shaken with the sample in a homogenizer to disrupt tissues and break open cells. Bead beating is very effective at releasing DNA from cells, but also causes DNA shearing, thus making it unsuitable for isolating high molecular weight DNA.

Proteinase K is a serine protease that cleaves peptide bonds in proteins. It is often added to lysis buffers because it is highly active in the presence of SDS, chaotropic salts, and elevated temperature (50°C), conditions that unfold proteins and make them more accessible for digestion. It also inactivates nucleases, making it is very useful for extracting high molecular weight DNA.

The Phenol–chloroform Method

Phenol has long been used to isolate RNA and DNA. It is an organic compound that can dissolve most proteins and separate them from DNA. Phenol is slightly water-soluble but has a higher specific gravity so a mixture of water and phenol can be separated by centrifugation into two phases. Because phenol is more soluble in chloroform than water, adding chloroform as an additional organic solvent further reduces the amount of phenol that would otherwise be carried over into the aqueous phase. DNA with an average size of 150 Kb or even much larger can be isolated by the phenol–chloroform method if performed carefully, partly due to reduced physical forces employed compared to column-based techniques.[2] It is also very effective at removing nucleases.

This method was once the standard approach for DNA extraction but has fallen out of favor because column-based methods are so much simpler. However, phenol–chloroform extraction of DNA is now seeing a resurgence for nanopore sequencing due to its effectiveness in generating long fragments. For instance, DNA having mean lengths of ~100kb and maximum lengths exceeding 1 megabase have been sequenced using this method

To perform phenol–chloroform purification, an equal volume of phenol or phenol–chloroform is added to the aqueous solution of lysed cells and mixed in a rotating flask until a fine emulsion forms.

The two phases are then separated by centrifugation, the aqueous phase on top and the denser organic phase below. At pH 8.0, DNA and RNA partition into the aqueous phase while proteins remain dissolved in the organic phase. Between the two phases, a white precipitate of proteins usually forms, which is known as the interphase. The aqueous phase containing the purified nucleic acids is saved, and the process is repeated a few times to ensure the complete removal of proteins before precipitating the DNA.

Ethanol Precipitation

Following the phenol–chloroform treatment of a cell lysate, DNA in the aqueous phase can be precipitated by adding a salt such as sodium or ammonium acetate along with enough ethanol to make a 70% ethanol solution. Ethanol is much less polar than water, and above a certain concentration it disrupts the hydration shells surrounding the DNA. This allows the salt cations to form ionic bonds with the phosphates of the nucleic acids, causing the DNA to precipitate. If DNA precipitates in large enough quantities, it resembles a spider-web with bubbles trapped in it (an effect caused by the outgassing of ethanol). In some cases, it can be hooked out in one piece or 'spooled' on a glass rod, but smaller quantities will form a pellet upon centrifugation. In both cases, the DNA needs to be thoroughly washed in 70% ethanol to remove residual salts before being resuspended in a low ionic concentration buffer at pH at 8.0.

Dialysis

Dialysis is a technique commonly used in protein purification but can also remove impurities from DNA. Because shear forces are minimized during dialysis, it is preferable to phenol–chloroform for isolating large DNA fragments. In molecular biology, dialysis separates molecules by their rate of diffusion through a semi-permeable membrane. Ions in solution will readily diffuse from areas of high concentration in the sample to areas of low concentration in the dialysis buffer until equilibrium is reached, but larger DNA molecules cannot pass through the membrane, and so they are retained. Dialysis is performed either by putting the sample inside dialysis tubing and submerging it in a large volume of buffer or, for smaller sample volumes, by carefully pipetting the sample onto a membrane floating on the buffer, a procedure called 'drop dialysis'. A useful side effect of this method is that DNA becomes concentrated over time as water moves out of the sample through the membrane. If higher concentrations are required, the dialysis can be performed over longer time intervals.

Megabase Sized DNA

Isolating megabase sized DNA requires significantly more time and effort than other techniques. In order to keep DNA molecules intact, they must be protected from hydrodynamic forces. A novel method that is being explored is to embed the cells in agarose blocks known as plugs.[3] The extraction procedure is then performed on the cells *in situ* by placing the cell-containing plugs in lysis buffer, digestion buffer, and wash buffer. DNA can be analyzed by inserting the plugs directly into a gel for pulsed-field gel electrophoresis (PFGE) or released from the gel using agarase enzyme which cleaves agarose into smaller subunits that can no longer remain in the gel state at room temperature. DNA released from agarose plugs requires further purification by dialysis, but this may not result in sufficiently high concentrations to be used for nanopore sequencing. Although plugs are promising, the technique requires further development, so we will go on to standard methods currently in use.

Input Requirements for Ultra-long Reads

One of the main impediments to generating ultra-long reads is having sufficient input material. For cells grown in culture this is not a problem, but for smaller samples another approach may be required. The approximate number of cells we have found are needed to generate ultra-long reads are given below for phenol–chloroform extractions (Table 7.1).

Table 7.1. Number of cells needed to isolate a minimum of 15 μg DNA for ultra-long library preparation.

Organism	Genome size	Number of cell required
E. coli	4.5×10^6	3.33×10^9
S. cerevisiae	1.2×10^7	1.25×10^9
C. elegans	1×10^8	1.5×10^8
H. sapiens	6×10^9	2.5×10^6

Quality Control and Fragment Size Assessment of DNA Samples

Performing the appropriate quality control (QC) on DNA extractions is vital to avoid disappointment when sequencing! The most commonly performed QC procedures are fragment size assessment, absorbance spectrometry, and fluorometric quantification. The TapeStation 2200 (Agilent) is a gel electrophoresis system used for fragment size assessment, although other instruments or conventional gel electrophoresis could also be used. One useful metric generated by the Agilent instrument analysis software is the DNA integrity number (DIN), which can be used to estimate the level of DNA degradation. A DNA sample with the majority of the DNA >60 Kb with little to no short fragments will have a DIN value of >9. If the sample shows a smear of short fragments, a sign of degradation, it will have a DIN value <1. For all MinION library types, a DIN value >9 is preferred because lower values will result in more short reads. A 0.4x SPRI cleanup (see Size selection with SPRI beads, below) is able to remove fragments below 1500 bp. A better solution is to begin with high-integrity DNA, then shear it down to the desired size, resulting in a tight fragment distribution with very few short fragments.

Absorbance Ratios

Another important metric for DNA quality assessment is the absorbance measured by a spectrophotometer such as the NanoDrop. This instrument measures the UV and visible light absorbance of the DNA sample, which permits quantification of both DNA and common impurities. The commonly used absorbance ratios for assessing DNA purity are the 260 nm/280 nm absorbance ratio and the 260/230 ratio. The 260/280 ratio is generally 1.8 for pure DNA. A lower value could indicate protein, phenol, or guanidine hydrochloride contamination. The 260/230 ratio is a secondary metric and is generally 2–2.2 for pure DNA, and a lower value may indicate phenol contamination. However, correct interpretation depends on the extraction method. If a spin column extraction kit is used, guanidine hydrochloride would be the most likely contaminant, whereas in phenol–chloroform extraction SDS or phenol contamination are more likely. Changes in sample pH can also affect 260/280 ratios, so the instrument should be blanked with the same buffer that the DNA is in. Nucleotide composition can also affect the measurement. For instance, AT-rich samples will have slightly higher 260/280 ratios than GC-rich samples.

Checking that absorbance ratios are consistent with pure DNA is an important QC step prior to nanopore sequencing. If there is a problem at this stage, it is best to repeat the DNA extraction to confirm that the ratios are repeatable. Figure 7.3 shows a Nanodrop

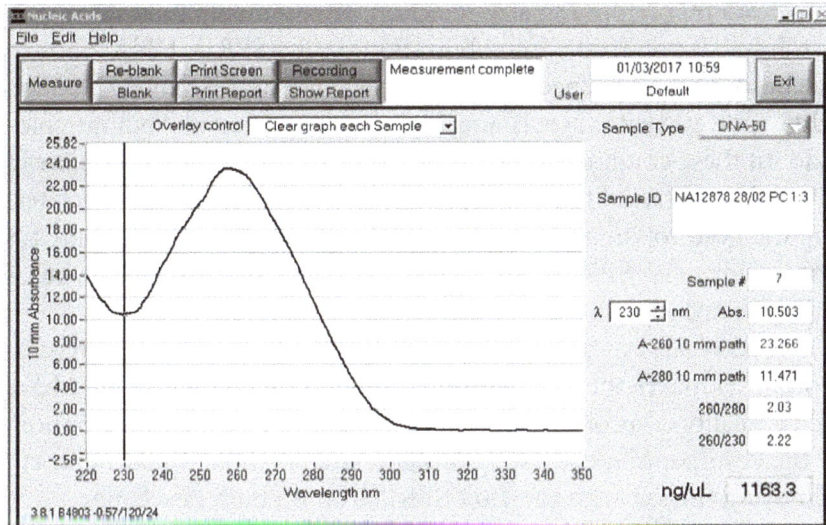

Figure 7.3. Absorbance spectra between 220 nm and 350 nm as measured by the NanoDrop instrument. This DNA sample was used to generate the ultra-long reads for the MinION human genome sequencing project. It was extracted from the NA12878 cell line using the phenol method.

spectrum for a high-quality DNA sample used for nanopore analysis, which has an even higher 260/280 ratio than expected for pure DNA. The Nanodrop instrument is useful for DNA purity assessment but less accurate than fluorometry for measuring DNA concentration.

Fluorescence spectroscopy is an important technique for DNA quantification. It relies on the fact that nucleic acid stains such as SYBR Green I fluoresce when intercalated in DNA. The level of fluorescence is proportional to DNA concentration which can be extrapolated from standards of known concentration. The Qubit (Life Technologies) is a convenient fluorescence spectrophotometer for single samples, and different kits are available for different sample types and concentration ranges. The most useful for preparing nanopore libraries is the dsDNA HS Assay (Life Technologies) which measures concentrations between 0.01 ng/μL and 100 ng/μL.

Size-selection with SPRI Beads

DNA extractions with evidence that short fragments are present can be improved by performing size selection. A commonly used technique is solid-phase reversible immobilization beads (SPRI). DNA binds to the beads in the presence of a buffer which contains a crowding agent, PEG (polyethylene glycol) and high concentration of sodium chloride. In these conditions, the DNA transitions to a condensed state in which it is attracted to the beads. Size selection is controlled by altering the bead to sample volume ratio, with ratios between 0.4 and 1.8 commonly used. SPRI is an easy way to remove short fragments but is only effective up to ~1,500 bp at the lowest ratio of 0.4.

SPRI beads can be used to clean up DNA prior to library preparation. This makes them useful for reworking DNA samples that have failed quality control by absorbance spectra or fragment distribution. If the absorbance spectra suggest salt contamination, it may be worth removing the salt with the 1.0× SPRI clean-up buffer exchange.

Many extraction kits use Tris-EDTA (TE) as elution buffer which contains 0.1 or 1 mM EDTA to protect DNA against nuclease activity. EDTA (ethylenediaminetetraacetic acid) is a chelating agent that sequesters metal ions like Mg^{++} that are cofactors of nuclease enzymes. However, such buffers should be used carefully because too high EDTA concentrations also inhibit the transposase enzyme used for nanopore library preparation.

Size Selection by Gel Electrophoresis

Agarose gel electrophoresis is used to separate DNA fragments by size. DNA is negatively charged and migrates toward the positive electrode when exposed to the electric field of an applied voltage. Typical gels are made with 0.5–2.0% (w/v) agarose, with lower percentage gels giving better resolution of long fragments because of their larger pore size. However, low-concentration agarose gels are fragile, and HMW (high molecular weight) DNA of different lengths tend to move together and cannot be resolved if a constant voltage is applied. PFGE on the other hand can separate fragments up to 10 Mb by applying voltage not just in one direction, but also at 60 degree angles from right to left. This causes large DNA molecules to move more slowly through the gel in a zigzag motion, while shorter molecules are separated because they move faster. The ability of PFGE to separate long fragments of DNA is exploited by instruments such as BluePippin and SageHLS (Sage Science) to perform size selection of genomic DNA. The most useful mode for nanopore sequencing is selecting the longest fragments in a DNA sample after g-TUBE (Covaris) or needle shearing, known as a high-pass size selection. Up to four samples can be size selected at once with the BluePippin 5-lane agarose cassette with the fifth lane used for the ladder. The DNA migrates through the gel until shorter, unwanted fragments have run past the collection channel. At this point, the anode is switched so the remaining longer fragments are eluted into buffer in the collection chamber. The switching point is determined by a DNA ladder running past a detector beneath the cartridge.

Repairing Damaged DNA

When sequenced read lengths do not match the known size distribution, the DNA may be damaged by single-stranded nicks caused by breakage of the phosphodiester bond between two adjacent bases in the strand. These can occur either by enzymatic activity or chemical damage. When the DNA strand is sequenced, nicks cause a premature termination of the sequencing read because there is no second strand to stabilize the nicked strand that is being driven through the nanopore. Single-strand breaks can be repaired using kits such as PreCR Repair Mix or FFPE DNA Repair Mix (New England Biosciences). These enzyme cocktails are designed to repair a variety of DNA damages, including single-strand breaks. Sequencing errors are reduced and read lengths improved, which can be especially important for old DNA samples. As an extreme example, ancient DNA (hundreds or thousands of years old) contains abasic sites, deaminated cytosine, oxidized bases, and nicks, all of which can be minimized by the FFPE DNA Repair Mix.

Handling and Storing HMW DNA

After expending so much time preparing high molecular weight DNA, a little extra care should be taken to ensure that work is not wasted. Although dsDNA is somewhat flexible, there is still a certain amount of stiffness due to the electrostatic repulsion between negatively charged phosphates.[4] This makes it vulnerable to double-strand breaks caused by hydrodynamic shear forces in moving fluids. Simply pipetting or vortexing a HMW DNA preparation can cause such breaks, but these can be minimized by pouring samples rather than pipetting them.

Maintaining high concentrations can also reduce shearing because high concentrations of DNA are more viscous. Keeping DNA in a condensed form by adding PEG or polyamines such as spermidine also reduces the likelihood of shearing.

When the DNA is to be stored, the preferred method is to resuspend it in elution buffer (EB; 10 mM Tris-HCl pH 8.0) or Tris-EDTA buffer (TE; 10 mM Tris-HCl pH 8.0, 1 mM EDTA). Nucleases are less active at pH 8, and the EDTA further protects against nuclease activity by chelating Mg^{++} ions. HMW DNA should always be stored at 5°C because freezing will result in physical shearing. DNA is stable for a year or more at this temperature if nuclease activity is absent.

References

1. Jain, M. *et al.* Nanopore sequencing and assembly of a human genome with ultra-long reads. *Nat. Biotechnol* **36**, 338–345. (2018).
2. Sambrook, J. & Russell, D.W. Isolation of high-molecular-weight DNA from mammalian cells using proteinase K and phenol. *Cold Spring Harb. Protoc.* **2006** doi: 10.1101/pdb.prot4036.
3. Schwartz, D.C. & Cantor, C.R. Separation of yeast chromosome-sized DNAs by pulsed field gradient gel electrophoresis. *Cell* **37**, 67–75 (1984).
4. Sambrook, J. & Russell, D.W. *Molecular Cloning: A Laboratory Manual* (Cold Spring Harbor Laboratory Press, Cold Spring Harbor, NY; 2001).

Chapter 8

Molecular Engineering DNA and RNA for Nanopore Sequencing

Andrew J. Heron

Most of the preparative methods used for extracting and processing genomic DNA (gDNA) result in randomly fragmented double-stranded molecules that have a mixture of blunt-ended and overhang ends. The molecules are not immediately suitable for nanopore sequencing in this form, and in general require the addition of three key elements: (1) a single-stranded end that enables efficient capture and threading of the DNA through the nanopore; (2) a method to load an enzyme motor to control the movement of the DNA through the nanopore; and (3) tethering elements to concentrate the DNA on the membrane to enable sequencing of small quantities of input sample. This chapter will outline how these elements have been brought together to prepare DNA as well as RNA for nanopore sequencing.

Capture and Translocation of Single-stranded DNA

Nanopore sequencing relies on a process by which an applied voltage draws single-stranded DNA (ssDNA) molecules through a narrow 1–2 nm constriction in a protein channel such as CsgG. The constriction in the pore is too narrow to pass genomic double-stranded DNA (dsDNA), which has a diameter of about 2 nm, so it is essential to find a way to efficiently control the delivery of the duplex DNA to enable passage of just one of the strands through the nanopore. Early studies demonstrated that synthetic double-stranded hairpins with blunt ends could be captured in the hemolysin nanopore entrance under an applied voltage. The duplex strands were separated by the force of the applied electric field, with one strand passing through the pore constriction, while the displaced strand was "unzipped" just before entry into the pore. However, the capture and translocation of the blunt-ended dsDNA was a highly inefficient process that required high concentrations of DNA in solution to observe a reasonable frequency of events.

The efficiency of DNA capture and translocation depends on many factors, including the shape and charge at the nanopore entrance, the buffer conditions, and the applied voltage. Capture and translocation can be improved by changing the structure at the ends of the dsDNA. Threading dsDNA first requires overcoming the energy barrier when unzipping the strongly hybridized duplex structure. By comparison, it is far easier to capture and thread ssDNA, since the ssDNA is already in the correct form to pass through the constriction. Researchers therefore began to attach synthetic single-stranded "leader" extensions to the ends of genomic dsDNA to improve the initial threading. After the leader is threaded, the force from the applied field acting on the translocated strand quickly overcomes the hybridization in the dsDNA section when that region encounters the constriction, causing it to unzip before entry into the pore.

The length, charge density, and flexibility of the leader all affect the threading efficiency, and over time the design and nature of the leader region has changed considerably. In early work, the leaders were most often single-stranded oligothymidine (oligoT) DNA, typically about 30 nucleotides long. These single-strand ends were also ideal for loading enzyme motors to control movement in the early development of the technology. However, as the chemistry has improved, leaders are now formed from non-polynucleotide charged polymers. These artificial extensions not only improve capture and threading, but also limit unwanted enzyme binding.

Enzyme Loading and Control of DNA Movement

The addition of leader sequences was an important step in the evolution of nanopore sequencing technology, but had little effect on translocation velocity. As described in Chapter 1, when a voltage greater than 100 mV is applied a ssDNA molecule will move through a nanopore at approximately a million bases per second, equivalent to one microsecond per base. This is far too fast to resolve individual bases, so over the years many approaches for controlling DNA translocation have been explored. Ultimately the most successful approach has been to repurpose nature's enzyme machines that have naturally evolved to process polynucleotides, and various studies have explored polymerases[1-3] and helicases[4,5] as motors to control movement with single-nucleotide resolution.

There were significant challenges in the initial search for enzyme motors capable of controlling DNA movement into a nanopore. In particular, many common polynucleotide enzymes were not tolerant of the salty buffer conditions required for nanopore analysis, or the forces applied by the voltage when strands were captured on the nanopore, which quickly dissociated the motor from the DNA. For example, one study[2] demonstrated that a polymerase could add bases to DNA held in the α-hemolysin nanopore, and in effect control the DNA position. However, while this approach demonstrated the feasibility of using an enzyme to control the position of DNA in a nanopore for sequencing, the enzyme could not directly control the DNA translocation because it quickly dissociated from the DNA under the force applied by the electric field. Two years later, Akeson and colleagues[3] discovered the first enzyme capable of remaining bound to DNA under the harsh nanopore sensing conditions for a useful duration. Phi29 DNA polymerase from the *Bacillus subtilis* phage phi29 is well known for its high processivity, replicating thousands of bases before falling off the DNA. Phi29 not only remained bound to DNA for several seconds on top of the nanopore, but was also able to control base-by-base movement of DNA through the nanopore. It could be used to control DNA movement either by acting as a slip-stick brake when DNA was dragged through the enzyme by the applied voltage, or by pulling the DNA back out of the pore when the enzyme was actively synthesizing DNA in the presence of its four nucleoside triphosphate substrates. Ultimately, both modes of movement were demonstrated to be sufficient to determine a DNA sequence.[6]

The phi29 polymerase was used in much of the early strand sequencing research and development effort. However, phi29 did not evolve to function as a motor on top of a nanopore, and it became clear the translocation of a DNA strand through a pore suffered from a variety of errors that inhibited effective *de novo* base calling. For example, random jumps a few bases forward caused nucleobases to be completed missed, and backward movements led to duplicate reads of the same base. The effort therefore focused on searching for other enzymatic motors capable of a more consistent stepwise movement. Many types and families of polynucleotide-binding enzymes were investigated, with a particular focus on helicases. The general function of helicases in a living cell is to translocate along one strand of dsDNA to unwind the double helix for replication (Chapter 5). Moreover, in many helicases the stepwise advancement along DNA is tightly coupled to adenosine triphosphate (ATP) hydrolysis, enabling some helicases to translocate along kilobases (kb) of DNA at speeds of more than 1,000 nucleotides per second before unbinding. These characteristics made them excellent candidates for nanopore sequencing.

Researchers at Oxford Nanopore first demonstrated that helicases are ideal motors for controlling DNA movement through nanopores.[5] After being optimized by protein engineering, they have the necessary processivity, speed, and step-wise movement that are essential for high-throughput and high accuracy nanopore sequencing. Much of the early proof of concept work focused on loading the enzymes from solution onto exposed ssDNA leaders that were ligated to genomic dsDNA. Early studies also demonstrated that there are many ways to employ helicase motors. For example, it is possible to use helicases to either feed the DNA into the pore, or pull the DNA back out of the pore, depending on which direction the enzyme moves along the DNA, and by controlling which end of the DNA was captured. The technology has developed considerably since then, and this chapter will describe how the enzymes are used in the current generation of nanopore sequencing chemistry.

Membrane Tethering of DNA Substrates, and the Effect on Sensitivity and Throughput

Once a strand has completed the sequencing process, the nanopore becomes unoccupied waiting to capture the next strand. This "open-pore" time between strands is wasted sequencing time; so to maximize the throughput of the system, the time between successive strands must be as short as possible, which requires getting as much DNA to the vicinity of the pore entrance as possible. To address this problem, the chemistry developed for the MinION makes use of hydrophobic tethers coupled to the DNA to concentrate the DNA near the membrane.[7] The sample DNA diffuses at random in the solution when it is first injected into the flow cell, but if the hydrophobic tether contacts the membrane, it "dissolves" in the hydrophobic hydrocarbon chains, effectively tethering the DNA to the membrane on whose surface it continues to diffuse, but now in only two dimensions. Concentrating the DNA onto the membrane in this manner, and relatively nearer to the pore, improves capture rates by 100–1000 times.

Genomic DNA samples are many kilobases in length, and usually tangle up into roughly spherical mass that may be hundreds to thousands of times larger than the size of the nanopore entrance. It is therefore not sufficient to simply bring this tangled mass near the pore, but even more important to deliver the appropriate end of the DNA to the nanopore entrance. The nanopore sequencing chemistry therefore places the hydrophobic tethers near the leaders to ensure optimal tethering of this critical region, which further improves capture efficiency by approximately 10 times. Since it is desirable that the sequenced fragments are as long as possible in most applications, an additional advantage of tethering only the ends of the DNA is that it does not matter how long is the rest of the attached genomic molecule. This reduces capture biases between shorter and longer genomic molecules, so that when sequencing samples with a mixture of fragment lengths, it is not significantly harder to capture a 50 kb fragment than a

1 kb fragment. This behavior does begin to break down when sequencing fragments larger than 100 kb because their sheer size means they behave less like simple molecules.

The enormous improvement in DNA capture rates provided by hydrophobic tethers reduces the recommended input requirements to amounts that are ideal for many real world sequencing applications. Typical inputs can be a microgram or less of starting genomic material per flow cell. For dsDNA with an average fragment length of 5 kb, this amounts to approximately 0.3 picomoles of DNA, a final concentration of approximately 4 nanomolar in a flow cell.

Nanopore Sequencing Adapters

Nanopore sequencing platforms such as the MinION are inexpensive, portable and easy to use, and produce real-time sequencing results. This makes them accessible to anyone for sequencing anywhere, including laboratories, field studies, and point-of-care settings, and are ideal for a diverse range of applications that require simplicity and rapid results. It is therefore essential that the methods for preparing the samples are also simple and fast and, where possible, without relying on expensive laboratory equipment or consumables. To address this, the leader strands, enzyme motors, and tethering elements are integrated into "Sequencing Adapters", which can be attached to genomic samples by a number of relatively fast means.[8] Furthermore, purified Sequencing Adapters ensure that leaders, enzymes, and tethers are specifically attached where required, thus providing a more robust and consistent method of treating the wide variety of genomic samples that will vary in composition and quality.

The Sequencing Adapters used for preparing genomic samples are individually tailored to the specific samples and methods, but they generally share a number of key features that are illustrated in Figure 8.1. On one end of the Sequencing Adapter is the leader extension that facilitates capture and threading through the nanopore. The leader is on the 5' end of the DNA Sequencing Adapters, and 3' end for direct RNA Sequencing Adapters. The leader is not composed of DNA, but instead is a highly charged flexible polymer that is optimized for capture

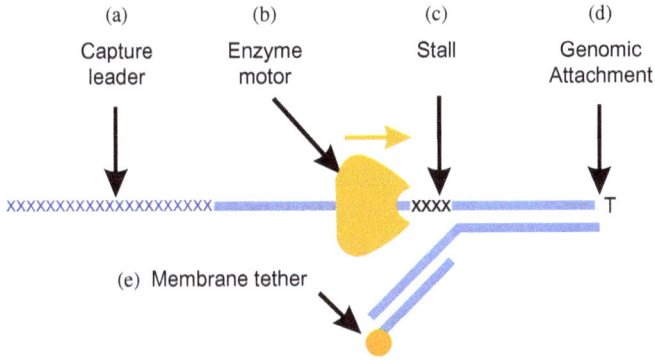

Figure 8.1. A typical Sequencing Adapter structure that is attached to genomic samples to prepare them for nanopore sequencing. The features include (a) a single-stranded polymeric leader with high negative charge density but no ability to bind polynucleotide-binding-proteins; (b) An enzyme binding site pre-loaded with a single enzyme motor. The arrow above the motor indicates the direction that the motor enzyme moves along the strand after it is physically pushed over; (c) the non-polynucleotide stalling section that prevents motor progression until it is captured in the nanopore; (d) A section to enable attachment to the genomic polynucleotide; (e) a side-arm structure to which the membrane tether is coupled.

and threading through the nanopore, thereby preventing hybridization to other polynucleotides in the sample and also preventing proteins such as the enzyme motors from binding to the polynucleotides. The opposite end of the adapters are typically double stranded, with the necessary structures and chemistry for attachment to the genomic sample.

At the core of the Sequencing Adapters are the appropriate pre-loaded enzyme motor for either DNA (currently E8 motor) or RNA (currently M1 motor) sequencing. Attaching adapters with pre-loaded motors provides a means of loading each polynucleotide with exactly the right number of enzyme motors for sequencing. The motor is also locked on the adapter around the DNA with a linker chemistry that prevents its unbinding.[9] Most natural enzymes are evolved to spontaneously bind and unbind from DNA, but in nanopore sequencing it is important to keep the enzyme bound, otherwise the strand will be prematurely lost during translocation. Locking the enzyme motor prevents it from unbinding until it encounters the end of the strand or a break in the strand. As a result, nanopore sequencing motors have essentially unlimited processivity, and there is no upper limit to the length of strands that can be sequenced as a single molecule on a nanopore system. A further advantage of locking the pre-loaded enzyme motor is that the Sequencing Adapters can be stored for many months in sequencing kits without loss of function.

The helicase motor enzymes used in nanopore sequencing are very powerful. They derive their energy from consumption of ATP, using the energy released during hydrolysis to adenosine diphosphate (ADP) to produce a power stroke that moves the enzyme along the DNA one base at a time. When activated, the motors are capable of not only unwinding double-stranded polynucleotide, but also pushing through polynucleotide tertiary structures such as g-quadruplexes, strongly bound proteins, and a wide range of damaged or modified polynucleotides. These are essential characteristics for the high-quality movement that underpins accurate base calling, and that are also needed for robustly dealing with diverse genomic samples. But once ATP is added to the system to begin sequencing, it is necessary to prevent enzyme progression until the complex is captured in the nanopore. The Sequencing Adapters therefore contain a stall region to prevent activated enzymes on uncaptured adapters in the *cis* chamber from progressing. [8] The stall is a non-DNA linker on which the enzyme cannot properly gain traction, thus temporarily halting its progression. When the leader is captured in the nanopore and the large diameter enzyme abuts against the nanopore into which it cannot fit, it is pushed over the stall region as the voltage bias drives the leader strand through the nanopore. After being pushed over the stall region, the enzyme re-engages with the DNA and continues to control the passage of the DNA into pore.

Finally, the Adapters also contain a region to bind the hydrophobic tether. Hydrophobic tethers can cause aggregation and other unwanted behaviors during library preparation, so the hydrophobic tether is typically a separate component that can be coupled, for example, by base pairing, to a side-arm in the final steps of the protocols.

Library Preparation and Sequencing Methods

A "library" is simply a collection of DNA strands that have been prepared for sequencing. The next sections cover some of the fundamental methods of attaching Sequencing Adapters to prepare both DNA and RNA libraries for nanopore sequencing. The library preparation methods for nanopore sequencing have been developed keeping in mind the core strengths of the technology surrounding portability, real-time sequencing, and long read lengths. They are therefore designed to be as simple and fast as possible, and to prepare genomic samples with long fragments. The methods are also designed to be modular and compatible with a wide repertoire of DNA handling methods, kits, and equipment. This section aims to provide a snapshot of the state of the current technology, but nanopore sequencing technology is still under active development, so it is reasonable to expect changes in future chemistries and methods.

There are many ways to attach Sequencing Adapters to DNA and RNA libraries to prepare them for nanopore sequencing. There are also various options for incorporating polymerase chain reaction (PCR) amplification to amplify some or all of the genomic sample, or barcodes to enable multiple samples to be combined into a single sequencing run. The flexibility of the preparation methods means they can be adapted to process a variety of samples, from viruses to microorganisms to the nuclear DNA of eukaryotes. For example, it is unlikely that it would be appropriate to prepare a small pathogen sample that requires rapid sequencing for species identification in a point-of-care scenario in the same way as a human sample for high-coverage genome assembly. This chapter will therefore provide guidance about the most appropriate library preparation methods, as well as a general overview of the two main methods for attaching Sequencing Adapters, by ligation or by transposition. The section will also cover various means to incorporate PCR amplification, barcodes, and $1D^2$ methods that enable higher sequencing accuracy. Finally, this section will also discuss the latest techniques by which RNA can be prepared for nanopore sequencing.

Ligation-based Library Preparation

Isolating and purifying DNA is described in Chapter 7, so only library preparation will be discussed here. Ligation is a common means of joining polynucleotide substrates together, and it is a straightforward and flexible means of attaching Sequencing Adapters or other adapters to gDNA to prepare it for nanopore sequencing. Ligation employs DNA ligase enzymes to join the ends of two DNA strands, which the enzyme facilitates by catalyzing a reaction between the 3'-hydroxy and 5'-phosphate termini to form a phosphodiester bond. A common DNA ligase that is recommended for nanopore library preparation is T4 DNA ligase, which is derived from the T4 bacteriophage that infects *E. coli* bacteria. T4 DNA ligase is available in various kits tailored for this purpose, such as the Blunt/TA Ligase Master Mix (M0367) provided by New England Biolabs.

Figure 8.2 illustrates the basic process by which Sequencing Adapters are attached to gDNA via ligation. In the basic protocol it takes approximately 60 min to prepare the DNA, and it involves a number

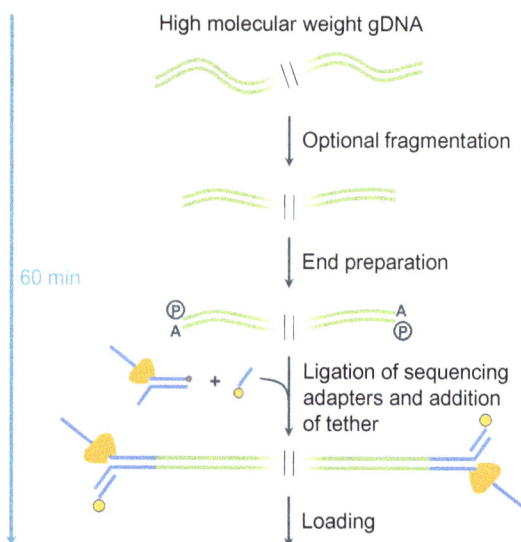

Figure 8.2. Process for preparing double-stranded gDNA for nanopore sequencing, involving end-preparation followed by ligation of Sequencing Adapters.

of straightforward handling steps and only minimal laboratory equipment. Although it will depend on the state of the input DNA, in most applications it is necessary to prepare the ends of the DNA to enable efficient ligation of the adapters. This is because gDNA can have a variety of end structures, including blunt ended and 5' or 3' overhangs depending on the source of the sample, and ligation is only optimal when the opposing ends of the adapter and gDNA are complementary. Since it is not practical to create Sequencing Adapters to match the multitude of different genomic ends, it is necessary to end-prepare gDNA to create homogenous ends. One convenient means of end-preparing gDNA is to use kits such as the NEBNext End repair/dA-tailing Module (kit E7546) provided by New England Biolabs. This kit prepares the DNA in a relatively simple single-tube process that takes approximately 10 min and involves one temperature cycling step. The kit contains a combination of exonucleases and polymerases that first digest or fill-in any ssDNA overhangs to yield blunted ends. In a second stage the 3' ends of the gDNA are extended with a single deoxy-adenosine, creating a "dA-tail". Sequencing Adapters and other adapters, with a single complementary 3' dT-tail, are available in the library preparation kits to match this type of end structure. The dA-tail/dT-tail complementarity between the gDNA and adapter creates a slightly stickier interaction that greatly improves ligation efficiency. The dA-tailing also prevents unwanted ligation products that would result from blunted end coupling, such as genomic-DNA to genomic-DNA ligations, or adapter to adapter ligations for example.

Although it is not strictly necessary, it is generally good practice to clean the gDNA after end-preparation and prior to ligation of adapters. The cleaning removes the enzymatic components that may interfere with the subsequent ligation step, and adjusts the buffers for optimal ligation efficiency. Cleaning is a relatively simple process when employing magnetic purification beads, such as those in the Agencourt AMPure XP bead washing kit, and takes approximately 15 min. The process essentially involves reversibly binding the DNA onto small beads, immobilizing the beads to the side of the tube with a magnet, washing away contaminants with clean buffer, and finally releasing the DNA from the beads in an elution buffer. Sequencing

Adapters can then be attached to the end-prepared (and optionally bead cleaned) gDNA by ligation. Ligation is a single-tube process in which the gDNA and Sequencing Adapter are combined with a DNA ligase and then incubated for about 10 min.

After ligation, the adapted DNA must be cleaned in another magnetic bead process as described above. The ligation-based library preparation protocols have been developed to generate maximum sequencing throughput, and this final cleaning step serves a few purposes to ensure the best possible sequencing performance. The primary reason for cleaning is to concentrate the gDNA sample and elute it into the final elution running buffer (ELB) in a state that is ready to be added to the flow cell for sequencing. ELB contains the high salt buffer required for sequencing, the ATP fuel and magnesium cofactor for the enzyme motor, and the hydrophobic membrane tether which couples to a side-arm on the Sequencing Adapter. This concentration and elution step means that the protocol can process variable amounts of input, for example from as little as 1 ng right through to many micrograms of DNA. While it is possible to start with less DNA, the recommended input requirements for ligation protocols are relatively high (1 μg of dsDNA for the R9.4 chemistry) so that the membranes are saturated with adapted DNA, which in turn leads to maximally occupying the pores during sequencing for the highest possible throughput. The bead cleaning also removes unligated adapter, which would otherwise compete with the genomic sample for nanopore sequencing time, and other contaminants that might block the pores. The bead cleaning does not remove long unligated genomic polynucleotides that have no adapter attached. However, these do not adversely affect the system because they lack hydrophobic tethers to concentrate them on the membrane and the leader structures required for efficient capture by the nanopore.

Transposase-based Library Preparation

Although preparing DNA by ligation is not a particularly long or difficult process, Oxford Nanopore have developed a "Rapid Sequencing" family of kits that employ transposase enzymes for exceptionally fast and simple attachment of the Sequencing Adapters. The protocols involve a few simple steps that prepare the DNA for sequencing in only 10 min, and require little or no laboratory equipment or additional kits. The Rapid Sequencing kits therefore complement the portable, real-time sequencing strengths of the MinION to prepare DNA for sequencing in applications where speed and simplicity are essential.

Transposase-based Attachment of Adapters to Genomic DNA

Transposases are enzymes that are able to cut and paste specific sequences called transposons into a different region of a genome, thus allowing organisms to rearrange their genomes for various biological purposes. Figure 8.3 outlines how the Rapid kits repurpose the transposases to simultaneously cleave the genomic dsDNA molecules and attach small adapters to the cleaved ends. The kits provide the transposase pre-loaded onto small transposon-like adapters, in a complex called a transposome. In this state, each transposome complex is primed to bind and react to gDNA, so the first step of the library preparation

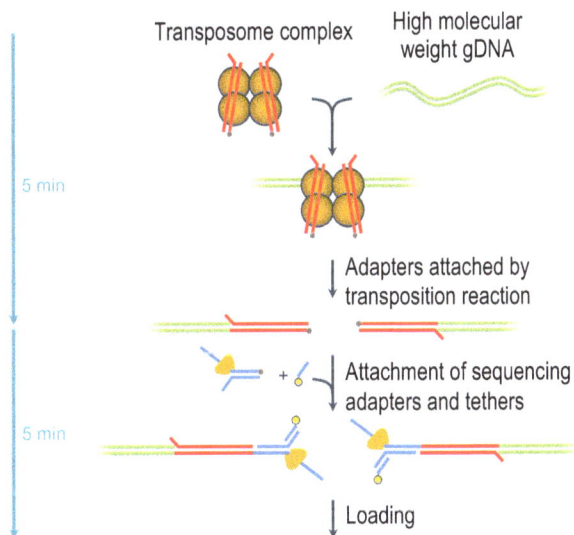

Figure 8.3. Process for preparing double-stranded gDNA for nanopore sequencing using transposase-based kits.

therefore simply involves adding transposome complex to the gDNA sample and incubating for approximately 1 min at 30°C. Upon mixing, the transposomes will bind randomly somewhere along the gDNA molecules and then proceed to cut the DNA, splitting the original dsDNA molecules into two new dsDNA fragments, and simultaneously attach one transposase-adapter to each of the new 5′ ends created.

The transposition is a single-turnover process. Each transposome binds, cuts, and attaches adapters only once, and will not react again. The entire process is also very quick, with the majority of transposome complexes fully reacting with the gDNA during the brief incubation. Since each transposition reaction results in two new smaller fragments, each with an adapter on only one end, transposition of an unmodified strand results in strands with only one end adapted. DNA strands with adapters on both ends are only generated from transpositions on strands that have already been reacted and contain at least one adapter, but at the cost of further fragmenting the strand. As a result, the extent of fragmentation is dependent on the molar ratio of transposome complex to DNA. Furthermore, since most transposomes will react during the incubation, the total number of tagged genomic ends is determined by the amount of transposase added, which in turn then determines the sequencing throughput.

After the brief transposition incubation, it is recommended to briefly heat the sample to about 75°C for 1 min. This could be in a heater if in a laboratory setting, but can also be any other source of heat in the 60–90°C range, for example even a cup of coffee in a location where no heater is available. The heating step serves to denature and unbind any residual transposase still attached to adapters or gDNA, which helps ensure maximal sequencing throughput as bound proteins can affect the enzyme motors or block the pore.

As illustrated in Figure 8.3, the basic Rapid sequencing kits typically use transposition to first attach a small transposase-adapter to the gDNA, rather than attaching the Sequencing Adapter directly. The final stage of preparing the DNA for sequencing therefore requires attaching the Sequencing Adapter to the transposase-adapted gDNA. This is discussed separately below, as the attachment method is employed in many library preparation kits.

Ligase-free Attachment of Sequencing Adapters

Early Rapid kits attached the Sequencing Adapter to the transposase-adapted gDNA by ligation, in a similar process to that described earlier. The ligation step was reasonably fast (about 10 min) and also very efficient because the ends of the adapters had matching sticky complementary overhangs to maximize the coupling efficiency. However, ligation attachment required an additional ligase kit. Newer Rapid kits instead use a ligation-free coupling chemistry to join the Sequencing Adapter to the transposase-adapted gDNA (Figure 8.3). The attachment method exploits the fact that the nanopore sequencing motors are locked onto the strands, and can therefore pass through a variety of non-DNA linking chemistry. This feature opens the option to link various DNA sections with conventional enzyme-free coupling chemistries that result in unnatural linkages that disrupt the DNA backbone. The coupling chemistry in the new Rapid kits uses reactive tags on the 5′-end of the transposase-adapter and the 3′-end of the Sequencing Adapter that spontaneously react to form a covalent bond when brought together. The tags on each adapter are different and cannot cross-react, so the coupling does not create any Sequencing Adapter–Sequencing Adapter or gDNA–gDNA species. Ligase-free coupling of the Sequencing Adapter is simpler and faster than ligation, and simply requires adding the Sequencing Adapter to the transposase adapted-gDNA and incubating for a few minutes at room temperature, Moreover, no additional kits or enzymes are required when using Rapid kits to prepare the DNA for sequencing.

Comparing the Ligation and Transposition Methods

The transposase-based and ligase-based sequencing methods are optimized for very different purposes, but can be applied broadly, so this section covers some of the key differences that may drive choice of method. These are summarized in Table 8.1 for the basic preparation protocols and kits. Both attachment methods are compatible with other optional processing steps, such as barcoding a sample for multiplexing, enriching for targets of interest using sequence capture,

Table 8.1. Comparison of the key differences between the ligation-based sequencing kits and the transposase-based Rapid kits.

	Ligation attachment	Transposition attachment
Typical preparation time	60 min	10 min
Input requirements	10 to greater than 1,000 ng Recommended: 1,000 ng of dsDNA	10 to a recommended maximum of 400 ng Recommended: 400 ng of high molecular weight dsDNA (>30 kb)
Read length	Same as input fragment length	Random shorter distribution vs input fragment length, dependent on amount of input material and transposome added
Expected yield (R9.4 chemistry)	High 3 Gb in 6 h 10–20 Gb in 48 h	Medium–High 2 Gb in 6 h 5–8 Gb in 48 h
Additional components required	End preparation kit Ligation kit Magnetic bead purification kit Thermocycler or heater	None (when using new ligation-free coupling) Thermocycler or heater (optional)

amplifying the sample using PCR, or $1D^2$ pairing for increased accuracy. There are advantages and disadvantages when including these additional processes, which are discussed later in the chapter.

The fundamental difference between the attachment methods is a balance of simplicity of library preparation against sequencing yield. The transposase-based methods are optimized for simplicity and speed, rather than obtaining maximum throughput. The simple processing steps, coupled with no requirement for other components or complex laboratory equipment, make then well suited to being taken out of the laboratory for in-field or point-of-care sequencing. Conversely, the ligation-based kits, with bead purification steps, are optimized for high-throughput by delivering large amounts of purified DNA. However, in the context of other sequencing technologies, many of which require dedicated and expensive lab-based equipment and 6 or more hours of library preparation, the ligation-based kits are still relatively fast and straightforward.

For R9.4 and R9.5 chemistry, the sequencing yields from DNA prepared with the transposase-based Rapid kits is typically lower than can be achieved for ligation kits. The difference in expected yield between the two methods is driven primarily by the limits of the final amount of DNA that can be added to the flow cell. The bead cleaning step in the ligation methods enables large quantities of DNA to be prepared, concentrating and purifying it into the final running buffer in the last step. It is therefore possible to prepare enough DNA to saturate the membrane when added to the nanopore flow cells, which in turn leads to maximal occupation of the pores during sequencing for the highest possible throughput. The cleaning also removes potential contaminants that might interfere with sequencing. While it is possible to bead clean transposase-adapted DNA, the basic Rapid protocols do not have a final bead cleaning and concentration step, which limits the amount of transposase complex that can be added. As a result, there may not be enough DNA to saturate the pores. However, the underlying nanopore sequencing chemistry is still under development, so future improvements should significantly reduce the amount of DNA or RNA required to saturate the pores and lower the recommended input requirements. It may eventually be possible to achieve similar yields from both transposase prepared and ligation prepared DNA.

Fragmentation is another key consideration when choosing which method to employ. Transposition inherently cuts the input DNA. While some of the original DNA may not be cut at all (and will not be sequenced), the sequences that are obtained will be the sequence of fragments that have been derived from the longer original fragments that have been cut at least once during attachment of the Sequencing Adapter. A proportion of these strands may derive from fragments have been cut many times depending on the relative amount of transposase to original DNA strands. This might entirely preclude the use of transposase-based methods in cases where it is important to maintain the integrity of the original strands needed, for example to obtain long continuous reads. In comparison, the ligation chemistry attaches adapters to the original DNA fragments without further fragmentation.

In many applications, when assembling a genome it is desirable that the sequencing reads are as long as possible. Since ligation attaches Sequencing Adapters without further fragmenting the gDNA, in an ideal case the final sequencing read lengths are generally more representative of the fragment lengths in the input sample. The ligation-based procedures are therefore the best to apply for obtaining long read lengths, and are generally suitable for DNA up to 200 kb in length as long as care is taken to handle the DNA very carefully. Application of a repair step prior to ligation is also strongly recommended to repair any damage that can prematurely terminate the sequencing strands. However, the effect that the entire library preparation process can have on the gDNA also needs to be considered. Very long DNA is fragile and prone to damage or breaks when handled, and the various handling and bead cleaning steps in the standard ligation-based protocols can fragment the input DNA. In comparison, the standard transposition-based methods require very little sample manipulation, and as a result very long sequencing reads can sometimes be easier to obtain from transposase methods despite the enzymatic fragmentation. Indeed, using the transposase-based kits, some users have sequenced single strands greater than 1 Mb in length on the MinION. Sequencing reads this long are unprecedented, but it is entirely feasible that with improved DNA handling the system will be capable of sequencing large mammalian chromosomes of >100,000,000 bases in a single pass.

Amplification by Polymerase Chain Reaction (PCR)

PCR is commonly used to amplify DNA for many sequencing applications. While there are several ways to implement PCR amplification in nanopore sequencing, PCR is not inherently required for nanopore sequencing. For most applications, it is recommended to sequence the original strands directly. This section will therefore briefly cover some of the key considerations when choosing to use PCR before covering the main methods of combining it into the library preparation methods.

The primary advantage of PCR amplification is the exponential increase in material it yields, so it is often essential when the available starting material is significantly less than the recommended input requirements. Amplification also provides a simple means of enriching a target of interest, for example a panel of genes in a large genome or a pathogen in a large background of human DNA, and many companies sell "amplicon" panels based on validated primers for enriching a range of specified targets. Another advantage of PCR amplification is the ease of including a barcode in the primers to enable convenient multiplexing of samples.

There are also some notable disadvantages to PCR amplification. PCR amplification complicates and lengthens the library preparation process and requires equipment that is capable of thermo-cycling the sample, which impacts applications where time and portability are important. PCR also limits the sequencing lengths that can be achieved because the polymerases can spontaneously dissociate from the DNA during synthesis. In most cases, it is not possible to get synthetic PCR products much longer than 10 kb, so the input DNA should be fragmented to less than 10 kb before amplification. The polymerases are also well known for creating biases and artefacts as a result of encountering particularly difficult sequences such as homopolymers, high GC or AT content or repeating sequence regions. This typically results in some regions with low coverage when assembling a genome, and artefacts or errors in the underlying sequences.

Finally, DNA damage and DNA modifications should be considered when using PCR. Damaged DNA is not properly amplified during PCR, so PCR is a useful means for enriching for good DNA in samples that contain large amounts of damaged DNA that might otherwise reduce sequencing yields or accuracy if sequenced without PCR. However, PCR creates copies of the original DNA using only unmodified deoxynucleotides, so any naturally occurring DNA modifications such as methylations are lost in amplified samples. Modified nucleobases play an important role in regulating and modifying biological function, for example in epigenetics, and are therefore under intensive investigation for a wide range of healthcare problems. One of the unique advantages of nanopore sequencing is the ability to directly detect a variety of naturally occurring nucleobase modifications, because they alter the current signal produced by unmodified nucleobases when they translocate through the nanopore. There are a number of advanced informatics tools available that detect these shifts, so PCR should not be used if it is desired to retain signals related to base modifications.

Ligation-based PCR Library Preparation

Ligation is a convenient means of attaching universal adapters with a known sequence for generically priming the ends of complex gDNA of unknown sequence for PCR amplification. The general process is illustrated in Figure 8.4. Like the ligation methods of attaching

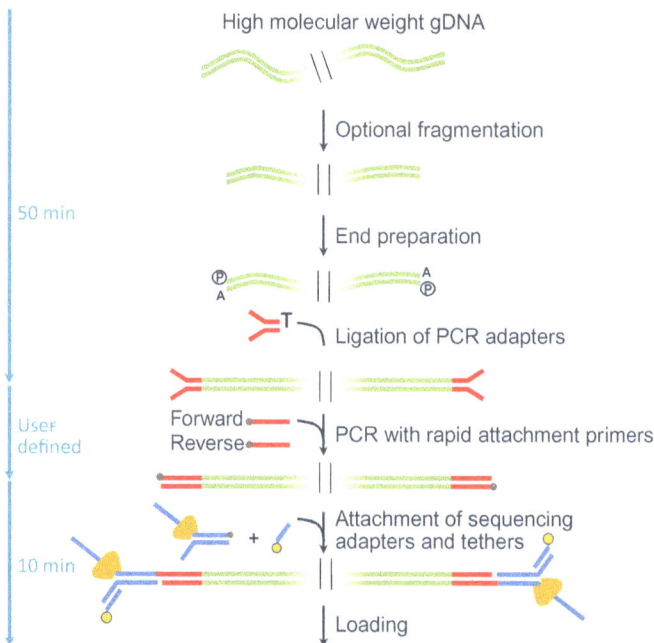

Figure 8.4.   Process for PCR amplifying and preparing DNA for nanopore sequencing, via ligation of universal primer binding adapters to the ends of the gDNA. The PCR amplification employs universal primers with 5′ coupling chemistry to enable ligase-free attachment of Sequencing Adapters in the final stage.

adapters described earlier, the dsDNA first has to be end-prepared, and then adapters that contain primer binding sites can be ligated to both ends of the dsDNA. The ligation process requires approximately 50 min of pre-PCR work. The PCR amplification process then employs "universal attachment primers" that are complementary to the known sequence in the attached adapters. The amount of time spent amplifying the sample is variable and depends on the number of cycles, the polymerase speed, and the template length, which needs to be tuned for each application. It is generally best to start with about 100 ng of gDNA, but it is possible to start with less, although it may then be necessary to increase the number of PCR cycles to further amplify the DNA.

Like the transposase-adapters described earlier, the universal primers provided in the kits have ligase-free 5′ coupling chemistry, so the matching Sequencing Adapters are simply attached by mixing them into the amplified and adapted DNA. Post-PCR work is therefore only about 10 min. Since the universal primer binding adapters are added to the ends of the original strands, the final fragment length after PCR amplification is approximately equal to the length of the DNA put into the PCR. However, PCR struggles to amplify fragments much longer than 10 kb due to the limited processivity of the polymerases. In samples with an average fragment length greater than 10 kb, the PCR will bias toward amplifying the shorter fragments, reducing the average fragment lengths, and reduce the amount of amplified material, so it is best to ensure that the original DNA is fragmented to less than 10 kb prior to amplification, using a Covaris g-TUBE for example.

Transposase-based PCR Library Preparation

Transposition is another convenient means of attaching universal adapters with a known sequence to complex gDNA to serve as priming sites for PCR amplification. This is illustrated in Figure 8.5. Like the transposase methods described earlier, kits are available containing a transposome complex formed from a transposase pre-loaded onto the primer binding adapters. When added to the gDNA, the

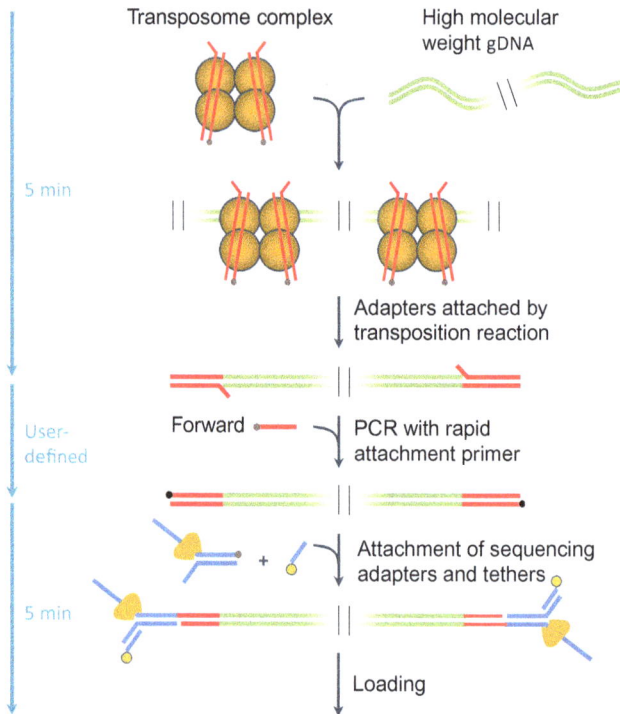

Figure 8.5. Process for preparing PCR amplified DNA for nanopore sequencing using a transposase to attach the primer binding adapters.

transposome cuts and attaches the primer binding adapter to the two new ends. Universal primers that are complementary to the attached adapter can then be used to PCR amplify the DNA. The universal primers provided in the kits also contain the 5′ coupling chemistry to facilitate simple ligation-free attachment of the matching Sequencing Adapters in the final step.

Transposition enables a far faster and simpler method of tagging the initial gDNA for PCR, reducing the hands-on preparation time in this stage to only 5 min. However, as with all transposase-based methods, the transposition inherently fragments the sample to some extent, dependent on the ratio of transposome complex to DNA molecules. Also, as described earlier, transposition does not optimally attach primers to both ends of the DNA, and only DNA with adapters on both ends will propagate through the amplification cycles. As a result of these issues, the fragment length produced after PCR is significantly shorter than that of the starting material, typically producing PCR products with a peak at around 2 kbs for most samples. Transposase based PCR amplification is therefore recommended for those who have very low amounts of starting material, necessitating amplification, but still require a faster library preparation process, but do not require long read lengths.

Preparing PCR Amplicons for Nanopore Sequencing

PCR provides a simple means of selectively amplifying target regions of interest to enrich them versus the unwanted background DNA. The general process employs primers that are designed to bind to either side of the region of interest to generate many copies of the target region during the PCR cycles. These amplified targets are called "amplicons", and many companies sell amplicon panels based on validated primers for enriching a range of targets. Amplicons formed by conventional primers yield blunt-ended double-strand DNA. These can be prepared for nanopore sequencing using the ligation methods for attaching adapters described earlier, thus entailing about 1 hour of post-PCR preparation. However, kits are also available to simplify the post-PCR preparation time by providing a means of incorporating the ligase-free coupling chemistry into the final amplicons. This enables ligase-free attachment of the Sequencing Adapters after PCR, which reduces the post-PCR preparation time to about 10 min. The process is illustrated in Figure 8.6(a).

Tailored nanopore sequencing amplicon kits are also available for very common applications. For example, the 16S kits provide primers that target the 16S rRNA genes of bacteria, which code for the 16S ribosomal RNAs in their ribosomes. 16S sequencing is by far the most common means of identifying bacteria and is widely used in applications such as the identification and tracking of pathogens. The 16S kit provides primers to target this approximately 1500 base pair region, and enrich over the rest of the genome through PCR amplification. This enables discovery of all the organisms in the sample without sequencing unnecessary regions of the genome, making

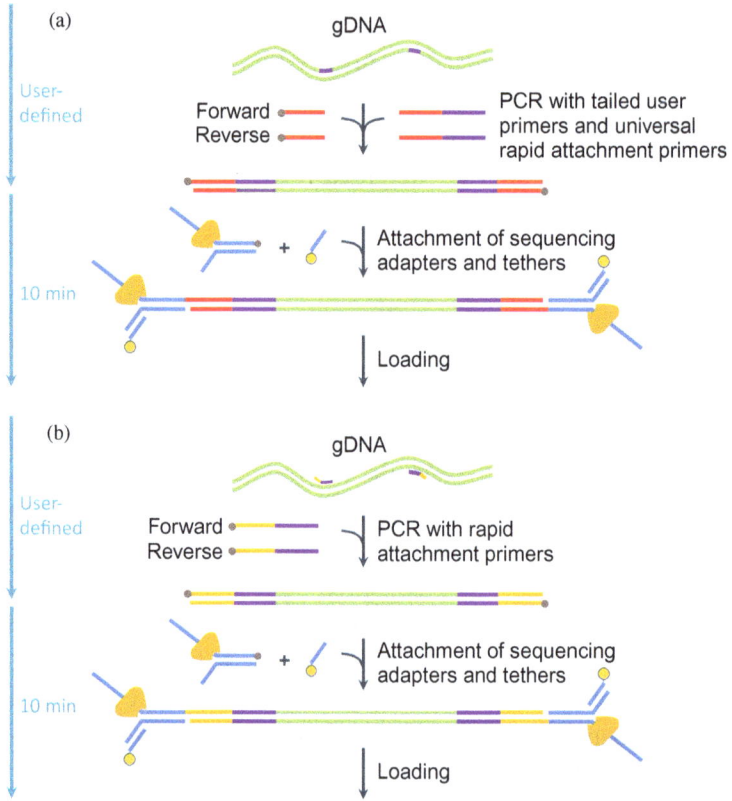

Figure 8.6. Methods for preparing amplicons for nanopore sequencing. (a) Preparing generic PCR amplicons for nanopore sequencing. Universal attachment primers combine with custom amplicon primers during PCR amplification to create amplicon products with 5′-coupling chemistry on the ends. The coupling chemistry enables simple post-PCR attachment of the Sequencing Adapters; (b) Preparing 16S ribosomal amplicons for nanopore sequencing using tailored primers with ligation-free coupling chemistry.

the identification quicker and more economical. The kit simplifies the post-PCR of attachment of the Sequencing Adapter by providing tailored primers with the ligase-free 5′ coupling chemistry (Figure 8.6(b)). Sequencing such a small region requires only a small amount of sequencing capacity to achieve high coverage, so the primers also contain barcodes to enable up to 12 samples to be pooled and sequenced together.

1D^2 Sequencing of Paired Template and Complement Strands

Although most genomic samples sequenced by nanopores are double stranded, the basic sequencing involves only one strand passing through the pore. This is therefore termed "1D sequencing". By convention, the first strand of duplex DNA captured and sequenced by the pore is called the Template strand. Its complementary annealed partner is called the Complement strand, and in normal 1D sequencing the Complement strand is unzipped above the pore while the Template strand translocates. Most of the time, the Complement diffuses away from the pore after it is fully unzipped from the Template and is never seen again because it is out-competed by the huge number of other strands in the vicinity of the pore. However, it is possible to encourage the Complement strand to immediately follow the Template strand through the pore by adding capture elements into the adapters that keep the strand in the vicinity of the pore. Sequencing a pair of Template and the Complement strands for a single dsDNA molecule is termed "1D^2 sequencing", and the process is illustrated in Figure 8.7.

1D^2 sequencing is a relatively new method for sequencing both strands of duplex DNA and is currently enabled by means of special modular adapter that can be incorporated into the basic ligation library preparation process described earlier in the chapter. As described earlier,

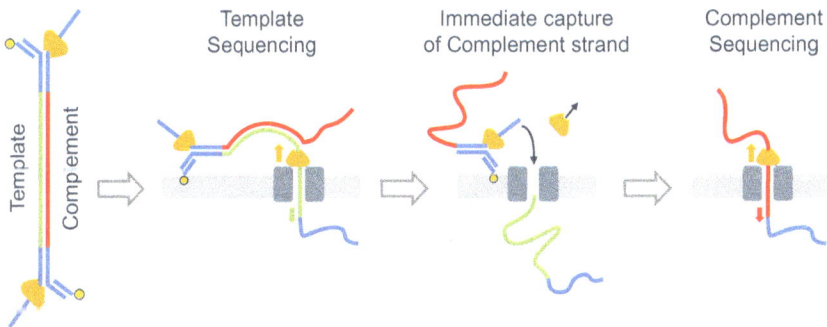

Figure 8.7. Schematic of strand translocation during 1D^2 sequencing where both the Template and Complement strands of the dsDNA are sequenced as separate strands.

the process first requires end-preparation of the gDNA to create dA-tailed DNA, followed by attachment of the intervening $1D^2$ adapter via ligation, and finally attachment of a Sequencing Adapter via ligation. $1D^2$ is currently not compatible with transposase prepared DNA, and also requires flow cells with the R9.5 pore type. However, $1D^2$ is still under development, so it will be extended to kits in the future and may not require a separate pore type.

The primary advantage of $1D^2$ is the ability to combine the information from both strands to learn more about the dsDNA molecule. For example, the base-calling algorithms can combine information from both the Template and Complement strands to overcome some of the single-molecule errors, and so yield a joint base call with a much higher accuracy for the dsDNA molecule. It is currently possible to achieve an average accuracy of greater than 97% using $1D^2$, and this may improve to >99% accuracy with future updates to the chemistry and base-calling algorithms. The increased accuracy for the dsDNA molecule can be very useful in a variety of applications where single-molecule accuracy is important, for example where sequences are rare and it is not possible to average out errors with high coverage of the same genomic region. The paired information can also be useful when examining heterogeneous samples such as human genomes which have different sequences in the same genes on either pair of a chromosome.

While the $1D^2$ adapters are able to encourage the Complement strand to follow the Template strand with very high efficiency, the entire process is not 100% efficient. Since $1D^2$ requires each strand to translocate separately using their own adapter with a pre-loaded enzyme motor, it is essential that adapters are attached to both ends of the gDNA. Attaching adapters by ligation is not a particularly efficient process, and typically achieves less than 90% attachment when linking strands at any one site. Because there are two ends to the dsDNA, two strands to be linked (the Template and the Complement), and two adapters to attach at each end (the $1D^2$ adapter and the Sequencing Adapter), these inefficiencies soon add up. For this reason, care should be taken to ensure optimal ligation conditions during library preparation. Damaged DNA can also affect $1D^2$ pairing, leading to premature termination of the strands. A DNA repair step is therefore highly recommended in cases where the DNA is damaged, or in long fragment samples where the likelihood of a nick increases with fragment length. With careful library preparation, taking these considerations into account, it is reasonable to expect about 60% of all strands to be paired. Therefore, from a $1D^2$ sequencing run that yields 10 Gb of 1D base calls, 6 Gb worth of data should be assigned as pairs, which can combined in the $1D^2$ base caller to produce 3 Gb of higher accuracy base calls.

RNA Sequencing

Although initially developed for DNA sequencing, it is also possible to sequence RNA on nanopore platforms. Like other sequencing technologies, the RNA can be sequenced indirectly from a complementary DNA (cDNA) copy of the RNA strand that is synthesized by reverse transcriptase enzymes. However, nanopores are just as capable of translocating RNA as DNA, and methods and enzyme motors are now available for direct nanopore sequencing of RNA strands.[10] Nanopore sequencing platforms such as the MinION are currently the only commercially available systems capable of direct RNA sequencing.

This section briefly covers how to prepare RNA for either cDNA or direct-RNA nanopore sequencing, and also discusses some of advantages that may determine which is the most suitable method to employ for the application. Table 8.2 summarizes some of the key differences between the methods.

Sequencing RNA Using cDNA

Reverse transcription enzymes (RNA-dependent DNA polymerases) are commonly used to create cDNA copies from RNA templates. The cDNA copy of the RNA can then be sequenced with conventional DNA methods to determine the original RNA nucleobase sequence. Nanopore sequencing of cDNA similarly involves sequencing the reverse transcribed DNA copy of the RNA strand, and Figure 8.8 illustrates the basic approach. The first step involves creating the cDNA copy by reverse transcription. The cDNA is polymerized from a poly-deoxyThymidine primer that anneals to the 3' poly-adenylated

Table 8.2. Comparison of cDNA and direct-RNA nanopore sequencing methods.

	cDNA	Direct-RNA
Preparation time	210 min	115 min
Typical Input requirement (R9.4)	250 ng RNA (polyA)	500 ng RNA (polyA)
Reverse transcription required	Yes	Optional
PCR required	No	No
Read length	Equal to cDNA length	Equal to RNA length
Read type produced	1D cDNA 1D^2 cDNA under development	1D RNA
Typical throughput	3–5 Gb	1 Gb
Typical number of reads	3–5 Million full-length	1 Million full-length

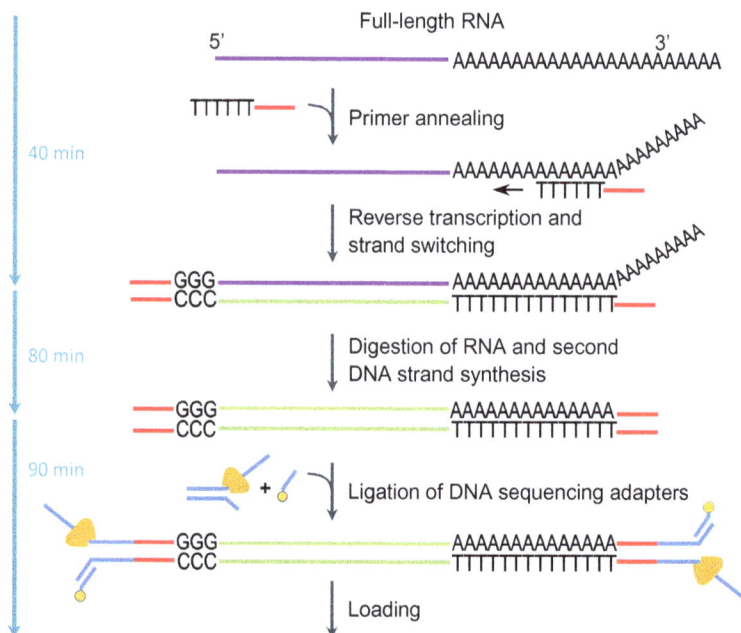

Figure 8.8. Preparing RNA for direct cDNA nanopore sequencing.

tail of the RNA. In a second step, the reverse transcriptase also creates a CCC overhang at the 3′ end of the new cDNA strand. This serves as an annealing site for a short complementary ribonucleotide sequence with GGG 3′ end, which acts as a template by which the enzyme can further extend the cDNA strand. This finally yields a cDNA copy with known short sequences at both ends.

The cDNA can be processed in a number of ways afterwards, and is treated much like regular DNA. As illustrated in Figure 8.8, it is possible to digest away the RNA template with RNAse enzymes, then fill in the top strand in a single polymerization step using a DNA polymerase to yield double-stranded cDNA. Alternatively, it is possible to further amplify the cDNA with multiple rounds of PCR amplification, which enriches for full-length cDNA strands that had primer binding sites on both ends. Various kits are available to support both direct cDNA and PCR cDNA methods. These also include primers that support options to add barcodes for multiplexing, $1D^2$ modules for higher accuracy, and coupling chemistry that enables simple ligation-free attachment of a DNA Sequencing Adapter.

As with DNA, PCR amplifies the original material, so the PCR-cDNA methods are recommended when there is a limited amount of input RNA, or where maximum throughput is important. Direct-cDNA sequencing is recommended for those with large amounts of input material (>250 ng for the R9.4 chemistry), and who wish to avoid any quantitative issues that may arise from PCR amplification biases.

The cDNA kits are provided with poly-dT primers to target any RNA with a poly-adenylated 3′-tail; however, it is possible to design primers for those who wish to sequence RNA without a poly-ade-nylated tail. While the poly-dT primers enable sequencing of any RNA with a poly adenylated 3′-tail, including viral RNA or RNA prepared with polyA-tailing kits, their primary purpose is to sequence eukaryotic messenger RNA (mRNA), which is naturally poly-adenylated prior to translation during gene expression. The study of gene expression via sequencing of mRNA is termed Transcriptomics, and is a field of growing importance as it enables researchers to properly analyze gene expression and function in response to environmental changes and disease. However, only about 5% of a eukaryotic cell's RNA content is mRNA, with the rest coming mostly from ribosomes, but also from tRNA, microRNA and non-coding RNA. In other sequencing technologies, it is therefore often necessary to pre-purify the sample to remove DNA (e.g. with DNAse) or unwanted types of RNA. A significant advantage of the library preparation required for nanopore sequencing is that the poly-dT primer naturally enriches for mRNA, and if subsequently amplifying via PCR the sample is further enriched with full-length transcripts as a result of the amplification cycles.

Unlike short-read sequencing technologies, the key advantage from a transcriptomic analysis point of view is that nanopore sequencing provides full-length cDNA reads. This is extremely important for fully understanding gene expression subtleties arising from RNA isoforms, splice variants, and fusion transcripts, all of which are implicated in many diseases.

Since cDNA is essentially sequencing of DNA, the cDNA kits employ the same motor protein used in the DNA kits (e.g. E8), running at approximately 500 nucleotides per second. Similar channel throughput and overall yields can be expected to that from DNA kits, depending on the amount of input material, and future updates to the DNA motor speed or the sensitivity will therefore also affect the throughput and input requirements for cDNA.

Direct RNA Sequencing

Figure 8.9 illustrates how the RNA itself can be prepared for direct nanopore sequencing. The first step of the process requires ligation of an adapter to the 3′ end of the RNA. Similar to cDNA sequencing, the kit is designed to prepare poly-adenylated RNA by employing a 3′-poly-dT overhang on the adapter; however, it is possible to design custom adapters to target any desired RNA sequence. This is followed by an optional (but recommended) reverse transcription step. The cDNA strand generated in the reverse transcription is not translocated or sequenced, but it significantly improves sequencing throughput by controlling problematic RNA secondary structure that can block the pores. In the last step, an RNA Sequencing Adapter is attached to the adapted RNA substrate by ligation. The RNA Sequencing Adapter contains a

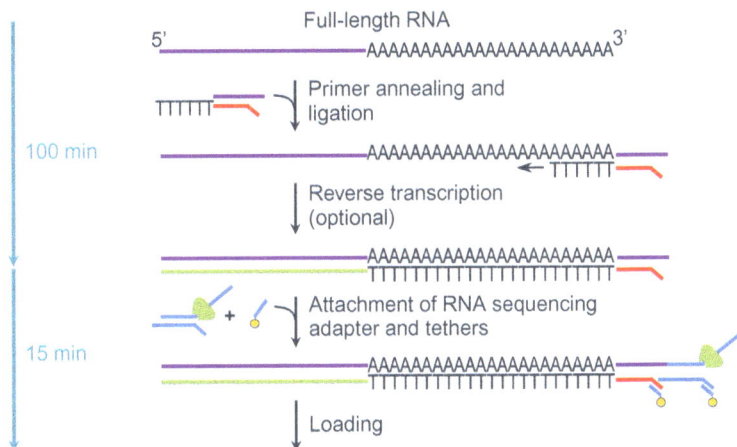

Figure 8.9. Process for preparing RNA for direct RNA sequencing on nanopores.

motor protein (e.g. M1) that is specifically designed to control RNA translocation through nanopores. In comparison to the E8 motor for cDNA/DNA, which runs at approximately 500 nucleotides per second and passes DNA into the pore 5'–3', the M1 motor passes the RNA into the pore 3'–5', currently at an average speed of 70 nucleotides per second. The substantially slower speed of the M1 motor protein results in significantly reduced throughput relative to E8-based cDNA sequencing. However, direct RNA nanopore sequencing is new and under continual development, so RNA sequencing throughput may in the future match that of DNA in most key metrics.

Like the nanopore cDNA methods, direct nanopore sequencing of RNA benefits from full-length reads that provide information about isoforms, splice variants, and fusion transcripts. Direct RNA sequencing also offers some unique advantages over cDNA. First, Direct RNA does not suffer from enzymatic synthesis biases that can affect quantitative cDNA sequencing in some applications. cDNA biases arise when the reverse transcriptase enzymes and the PCR amplification polymerases encounter sequences that are incorrectly copied, either due to base content or strong secondary structure. Difficult sequences can result in a variety of biases, including copy artefacts, and missing or truncated cDNA copies. In comparison, with a few rare exceptions, Direct RNA sequencing reads the full-length strand to its true end. Second, a key advantage of Direct RNA sequencing is that it enables detection of modified RNA bases, information that is lost when reverse transcribing is used to prepare cDNA copies. Modified RNA bases play an important role in regulating and modifying RNA function in biological systems, and are under intense investigation for a wide range of healthcare problems.

Maximizing Sequencing Throughput

Nanopore sequencing throughput, which is the amount of data per unit time, is primarily dependent on the number of active sequencing channels and on the speed of the enzyme motor, which has improved from 30 nucleotides per second in 2014 to approximately 500 nucleotides per second in the R9.4/E8 chemistry. The speed updates have driven a dramatic increase in the sequencing yields that can be obtained from a MinION. However, the throughput and ultimate yield can vary significantly depending on the input sample, so this section briefly covers some of the aspects to consider when preparing sequencing libraries.

Nanopore Blockages

In an ideal situation, the throughput of a nanopore is determined by the proportion of time spent in strand sequencing versus the unoccupied open-pore time waiting to capture another strand. However, in practice a significant amount of nanopore time can be wasted on unwanted captures and blocking, ultimately reducing the yields. Some blockages can occur from incorrect capture of sequencing components, but many blockages are also the result of capturing impurities carried through from the input sample. Where possible, it is therefore best practice to take steps to ensure the input DNA is as clean as possible.

It is however not always practical for the input DNA to be perfectly clean when sequencing complex samples, so to ensure that the nanopore system provides robust sequencing performance for diverse applications it has been optimized to tolerate unwanted blockades. Furthermore, the software running the MinION continuously monitors each pore, detecting and attempting to unblock any pores that are blocked for more than a few seconds by briefly reversing the applied voltage. The process is very effective, ejecting the vast majority of blockages within seconds of their occurrence. In a normal run, unwanted captures and blocking should not account for more than 20% of the total pore time. However, some blocks cannot be unblocked, so it is normal to see a small amount of pore loss over the course of a sequencing run due to permanent blocking. For the R9.4 chemistry, pore loss should be less than 20% of viable pores every 8 hours.

Maximizing Pore Occupancy

After excluding unwanted blockages, the remainder of the time is spent either sequencing a strand or in an open-pore state awaiting capture of next strand. For any given time period, if the total amount of time spent sequencing strands is t_{strand}, and the total amount of time in open-pore is $t_{open-pore}$, then the average percentage of time spent with the nanopore occupied by a traversing DNA strand will be $t_{occupancy} = 100\% * t_{strand}/(t_{strand} + t_{open-pore})$.

The time during which the pore is occupied by a traversing DNA strand will be directly proportional to the average length of the DNA fragments and the speed of the enzyme. The open-pore time between strands, $t_{open-pore}$, is approximately proportional to the concentration of adapted DNA ends added and the sensitivity of the chemistry.

Optimal flow cell sequencing throughput is achieved by maximizing the pore occupancy. So in an ideal case where there is no blocking, throughput is directly proportional to average pore occupancy of all the active sequencing pores. Figure 8.10 shows examples of a pore maximally filled under high occupancy conditions from a high-throughput flow cell versus a low occupancy pore with excessive open-pore time from a low throughput flow cell. Although it is possible to add less sample, for the R9.4 chemistry a starting input of 1 microgram DNA will achieve an optimal pore occupancy of about 90%.

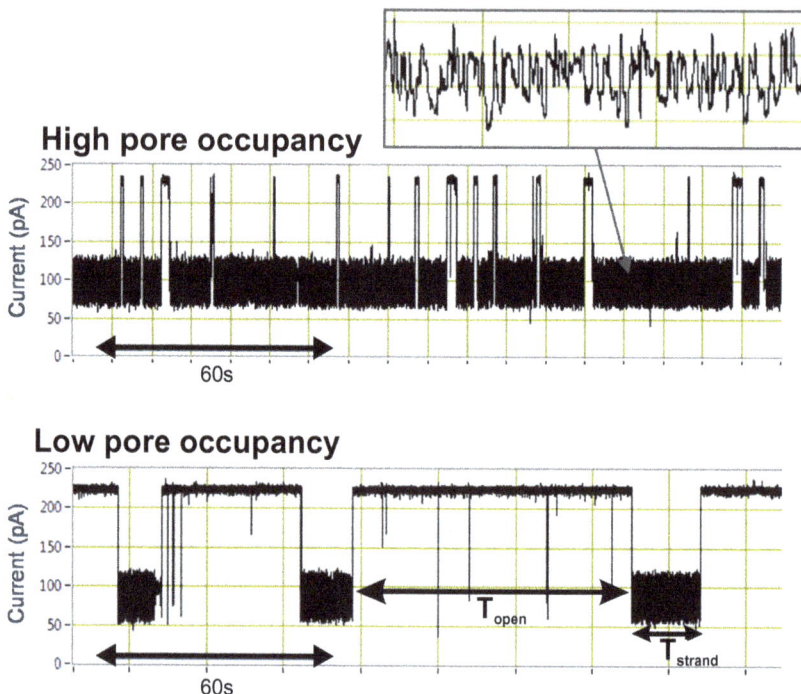

Figure 8.10. Examples of sequencing pores under high-throughput, high strand occupancy conditions (top) and low-throughput, low strand occupancy conditions (bottom). The graphs show the current versus time signals for single pores. The high current level is from the unoccupied pore, and the current drops when a strand is captured. The insert shows a zoomed section of the base-dependent changes in the current from a translocating DNA strand.

In general it is always better to add more DNA to the system, but it is important to note that the throughput will not scale linearly with amount of input material. This is because the concentration of material added only affects the open-pore time spent waiting for the next strand, with an approximately linear relationship, not the time spent in strand sequencing. Therefore, in a low occupancy regime where there is much more open-pore time than strand time (e.g. <50 ng), reducing the open-pore time by doubling the amount of sample added does almost double the throughput. However, in a high occupancy regime (e.g. >500 ng), reducing the open-pore time by doubling the amount of sample added makes little difference to the overall throughput because there is so little open-pore time. Figure 8.11 illustrates how pore occupancy and throughput vary as a function of input amount.

The recommended input guidelines are set to ensure the best possible sequencing throughput, but there is in fact no minimum amount of material that can be added to a flow cell for sequencing. The nanopore sequencers are single-molecule systems after all, so adding fewer molecules simply means waiting longer between successive strands for the pore to capture the next molecule. This can be very important for those who have much less than the recommended input requirements, but also do not require much sequencing yield. For example, users have demonstrated that just a single long nanopore sequencing read can be sufficient to identify a bacterial species. It is therefore possible to add 10 ng of input material, or even less if desired, but pore occupancy and throughput will drop as illustrated in Figure 8.10, and it will take longer to acquire the same sequencing yield. However, the sensitivity of the nanopore chemistry is under continual development,

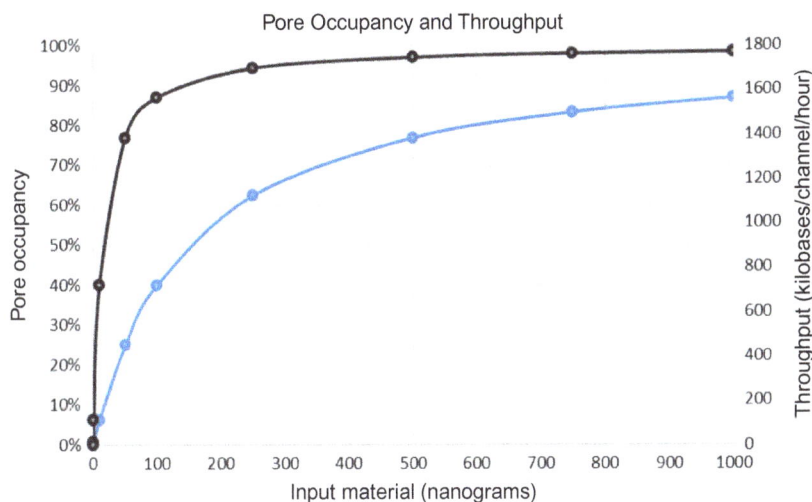

Figure 8.11. The dependence of pore occupancy and throughput on the amount of DNA input material. The blue line shows the approximate pore occupancy and throughput that can be achieved on well-adapted DNA (R9.4 chemistry, E8 motor running at ~500 b/s) versus amount of input material. The black line shows the improvement in pore occupancy and throughput versus input material that can be expected from a 10x boost in the sensitivity of the sequencing chemistry.

which means that smaller input amounts will be required to achieve optimal pore occupancy. Figure 8.11 illustrates how improvements in sensitivity can impact the input requirements.

It might be expected that throughput depends on fragment length, since the number of DNA ends will decrease proportionately with increasing fragment length for the same mass of starting material. However, because the average throughput depends on both time in strand occupancy and time between strands, the decrease in concentration of ends is balanced by the increased time spent sequencing a strand. For example, for the same 1 microgram input, a sample with 10 kb fragments would have 10 times fewer ends than a sample with 1 kb fragments, so would have 10 times longer open-pore time between each strand. However, this would be offset by spending 10 times longer sequencing each strand. This simple generalization, which fundamentally assumes shorter strands behave much the same as longer strands during a sequencing run, holds for fragment lengths in the 1–20 kb range. A consequence of this behavior is that there is no specific strand bias in mixed fragment length samples, so the length distribution of sequenced strands lengths matches that measured by other DNA quantification methods. For much smaller or much longer fragment lengths, these assumptions can break down and are further complicated by differences in attaching adapters and nicks in longer strands prematurely terminating strands.

Barcoding and Multiplexing

For many applications, it is cost-effective to pool multiple libraries for collective sequencing on a single nanopore flow cell. This process, called multiplexing, enables more efficient use of flow cell sequencing potential. For example, in low input cases such as that discussed in the previous section, when there is insufficient material in a single sample to saturate the pore occupancy, adding multiple samples will fill more of the open-pore time and make more efficient use of the pore time.

Alternatively, even if a single sample fills all the available pore time, if an application requires only a portion of the sequencing yield that the flow cell is capable of achieving (e.g. 1 Gb of sequencing is required, but the flow cell can yield up to 20 Gb), then further samples can be added to the flow cell to use the rest of its sequencing potential. This can be achieved by pooling multiple samples and multiplexing them in a single sequencing run. Multiplexing in this manner makes better use of the flow cell potential over its full lifetime, but the throughput *per sample* would drop in proportion to the number of samples added as all the samples compete for nanopore time during the longer run that is employed to acquire the desired coverage in each sample. If it is not desirable to wait for enough samples to multiplex, for example if there are time pressures to obtain the data from single samples as quickly as possible, then it is also possible to run multiple samples as a series of separate sequencing runs, washing out old samples between each successive run. This is discussed further in the next section.

Multiplexing multiple samples on a single flow cell requires a unique barcode to disambiguate the sample origin for any specific single-molecule read. Barcoding is achieved by attaching short unique sequences, typically <40 bases, to the individual genomic samples during library preparation prior to pooling. Barcoding modules are available for almost all DNA and RNA sequencing kits. The inclusion of the barcode sequence depends on the kit and involves either direct inclusion in the Sequencing Adapters, ligation of a separate barcode adapter, or primers that allow incorporation during PCR amplification (Figure 8.12). The various barcode modules include validated sequences that are detected and assigned by the appropriate barcoding analysis workflows, supporting multiplexing of up to 96 samples. Some combinatorial barcode modules will support multiplexing of up to approximately 10,000 samples, and details are available to enable creation of custom barcodes and analyses to suit further application requirements.

Reusing Flow Cells and Recovering Samples

A key feature of nanopore sequencing flow cells is that they are not designed to be used once and then discarded or refurbished. Rather, they have a limited lifetime that depends on the number of channels with intact membranes and viable nanopores, and on sufficient amounts of the ferri/ferrocyanide mediator solution in their imme-

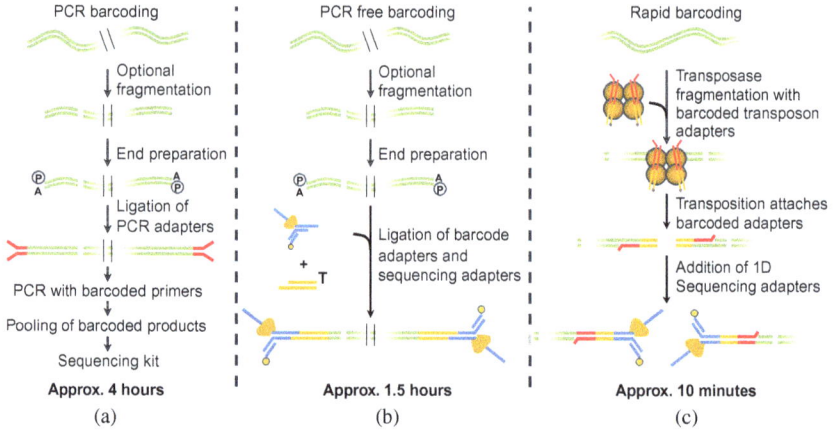

Figure 8.12. Barcodes for sequencing multiple samples together can be attached to the genomic sample by PCR primers during PCR amplification (a), by ligating a separate barcode adapter (b), or by using barcoded transposome adapters (c).

diate vicinity of the electrodes to pass current through the system. Flow cells can be stopped and restarted as desired, with long-term storage possible between uses, and therefore can be reused. Furthermore, if a sample has been run long enough to yield sufficient sequencing data and there are still adequate numbers of active pores, then the sample can be removed to enable subsequent sequencing of additional samples.

The use of the hydrophobic tethers that bind the DNA to the membrane means that the genomic sample cannot simply be removed from the system by flushing buffer through the flow cell. Wash kits containing a chemistry to de-tether the genomic molecules are therefore available to release the DNA back into solution. Once released into solution, the sample can either be washed through to the waste section of the flow cell by flushing through fresh buffer, or recovered by pipetting the solution back out of the flow cell. Figure 8.13 shows an example of how a flow cell can be reused to sequentially sequence 8 separate barcoded samples, with overnight storage between each sequencing run. The wash kit process is not 100% efficient, so traces of

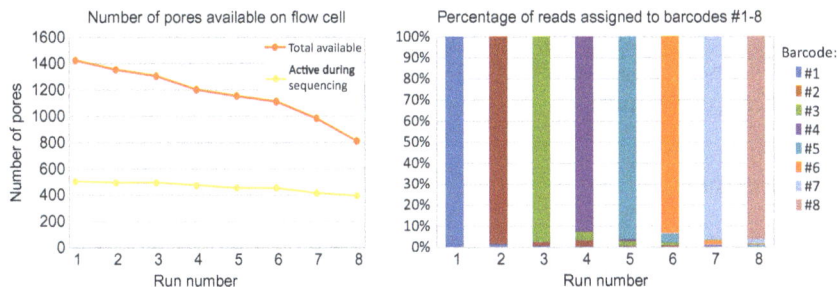

Figure 8.13. Sequence successive independent samples. Example showing how nanopore sequencing flow cells can be stopped, stored, and restarted to separately sequence successive independent samples, using wash kits that enable the old sample to be removed or recovered between sequencing runs. This example shows 8 successive sequencing runs on one flow cell, sequencing 8 samples with unique barcodes numbered 1–8. At the end of each run, the old sample was removed with a wash kit and the flow cell was stored overnight. (Left panel) Although there is a gradual loss of the total number of available pores on the flow cell over the course of the runs, the number available for active use remained at about 400–500 in each run. (Right panel) The percentage of reads aligned to each barcode for each sequencing show that very little of the previous samples remain in successive runs.

the previous sample can remain. It is therefore best practice to barcode the sequential samples in cases where it is necessary to unambiguously discriminate between the new samples and the trace background.

The option to recover the sample by pipetting the DNA solution back out of the flow cell is another key feature of nanopore sequencing, and can be important when the sample is precious or limited. The nanopores only consume a very small fraction of the sample, typically less than 0.001%, by permanently translocating the strands into the wells of the chip, so the majority of the DNA can be recovered. To ensure that most of the sample is recovered, it is recommended that a run-volume of solution be removed prior to the wash kit process and collected along with that recovered after using the wash kit. This ensures recovery of both the tethered DNA and the DNA free in solution. The balance between how much is tethered and how much remains free in solution depends on how well the adapters were attached to the genomic sample and whether the DNA was added in excess to saturate the membrane.

After the DNA is recovered, it can be processed and purified with common methods for later use in further sequencing experiments or other analytical techniques.

References

1. Benner, S. *et al.* Sequence-specific detection of individual DNA polymerase complexes in real time using a nanopore. *Nat. Nanotechnol.* **2**, 718–724 (2007).
2. Cockroft, S.L., Chu, J., Amorin, M. & Ghadiri, M.R. A single-molecule nanopore device detects DNA polymerase activity with single-nucleotide resolution. *J. Am. Chem. Soc.* **130**, 818–820 (2008).
3. Lieberman, K.R. *et al.* Processive replication of single DNA molecules in a nanopore catalyzed by phi29 DNA polymerase. *J. Am. Chem. Soc.* **132**, 17961–17972 (2010).
4. Astier, Y., Kainov, D.E., Bayley, H., Tuma, R. & Howorka, S. Stochastic detection of motor protein-RNA complexes by single-channel current recording. *ChemPhysChem* **8**, 2189–2194 (2007).
5. Moysey, R. & Heron, A.J. Enzyme method. WO Patent 2013057495 (2011).
6. Cherf, G.M. *et al.* Automated forward and reverse ratcheting of DNA in a nanopore at 5-Å precision. *Nat. Biotechnol.* **30**, 344–348 (2012).
7. Clarke, J., White, J., Milton, J. & Brown, C. Coupling method. WO Patent 2012164270 (2011).
8. Heron, A.J. *et al.* Enzyme stalling method. WO Patent 2014135838 (2013).
9. Heron, A. *et al.* Modified helicases. WO Patent 2014013260 (2012).
10. Garalde, D.R. *et al.* Highly parallel direct RNA sequencing on an array of nanopores. *Nat. Meth.* (2018) **15**, 201–206.

Chapter 9

Bioinformatic Analysis of Nanopore Data

Miten Jain

Earlier chapters described how a MinION can determine base sequences of single-stranded nucleic acid molecules (DNA or RNA) as they are translocated through a nanopore by an applied potential. The translocation rate is controlled by a helicase enzyme that rachets the strand through the pore at ~450 bases per second (b/s). The potential, or voltage applied across the membrane, produces an ionic current that flows through the pore along with the nucleic acid. The current is modulated by combinations of nucleobases in the nucleic acid called *k*mers, and each *k*mer has a specific effect on the ionic current. The ionic current modulations can be related to specific *k*mers which allows the sequence of bases in the nucleic acid to be determined.

The MinION provides the sequence data as a file that must be interpreted by specialized software, then analyzed by bioinformatic programs that allow users to align the sequences to a reference sequence and to determine the accuracy of the sequence that is read. This chapter describes how the software can be downloaded and used and includes an exercise that lets new users learn the basic steps.

Because progressing through much of the information and exercises in this chapter requires opening network sites when working at a computer running Mac OS X or Linux, a copy of this chapter has been made available at https://www.worldscientific.com/worldscibooks/ 10.1142/10995#t=suppl to facilitate using links or copying and pasting to open network sites.

Basecalling

The ionic current changes arising from a DNA or RNA molecule translocating through the nanopore are interpreted by a process called basecalling. This process identifies nucleotide combinations from a lookup table established by matching the ionic currents to the sequence of bases in a known DNA such as the M13 virus with a sequence of 7 kilobases (kb). This comparison makes it possible to assign different ionic current signatures down to the *k*mer level. The *k*mers used in MinION DNA sequencing are 6-mers, and 5-mers in RNA sequencing.

Once a reliable lookup table is available, it can be used for *de novo* basecalling. For instance, the ionic current modulations can be recorded when an unknown nucleic acid molecule is translocated through a nanopore. The next step is to search for the *k*mers in the table that are likely to represent those changes in ionic currents. The result is a DNA or RNA sequence that corresponds to the molecule that gave rise to the ionic current modulation. The newer versions of MinKNOW, the software that runs a MinION, includes basecalling within itself. This allows it to directly yield sequence file(s) in FASTQ format from a MinION run. When you finish a MinION run, you can go and find these sequence files in the destination that was specified for files to be written to earlier (when installing MinKNOW software). You can then take the MinION output files (containing the FASTQ sequences) and perform all sequence-based analyses such as alignment, statistics calculations, and assembly described in this chapter. The database that matches different *k*mers to their associated ionic current patterns uses algorithms such as a hidden Markov model (HMM; https://en.wikipedia.org/wiki/Hidden_Markov_model) or a recurrent neural network (RNN; https://en.wikipedia.org/wiki/ Recurrent_neural_network). The present MinION basecalling uses RNNs, and an example can be viewed at https://github.com/nanoporetech/scrappie.

Basic Bioinformatics Analyses

Although the MinION provides information about base sequences in DNA or RNA, the significance of the sequences cannot be understood until they are analyzed by bioinformatics techniques. There are two broad categories of analysis: (1) when a reference sequence is available, the analyses performed usually start with an alignment to that sequence; and (2) when this information is absent, the analyses performed are *de novo*, and usually start with an assembly.

Practice Exercise

What follows below is an exercise to demonstrate basic bioinformatics approaches that can be applied to MinION results. The exercise assumes that a reference sequence for a known genome is available, in this case from a bacterium called *E. coli*, and the exercise will align a MinION sequence to the reference sequence. The brief glossary and links below define terminology used in the instructions.

Glossary

1. Python — Programming language (https://en.wikipedia.org/wiki/Python_(programming_language))
2. Anaconda — Python and Python package manager (https://www.anaconda.com/download/#macos)
3. *E. coli* — *Escherichia coli*, a common bacterium that contains 4.5 million bases in its reference sequence (https://en.wikipedia.org/wiki/Escherichia_coli)
4. SAM — Sequence Alignment Map, an alignment file format (https://en.wikipedia.org/wiki/SAM_(file_format))
5. BAM — Binary Alignment Map, an alignment file format that is readable by a computer (https://en.wikipedia.org/wiki/SAM_(file_format))
6. FASTA — A sequence file format, that contains a unique identifier for every sequence along with the sequence itself (https://en.wikipedia.org/wiki/FASTA_format)

7. FASTQ — Also a sequence file format, that contains more information that a FASTA format (https://en.wikipedia.org/wiki/FASTQ_format)
8. Alignment — The process of measuring how similar two sequences are, such as a read sequence vs. a reference sequence
9. Assembly — The process of putting together a large sequence using smaller sequence
10. QC — Quality control. This refers to the accuracy of the base-called sequences
11. Pip — A type of Python package manager
12. Git — A Python library that allows downloading and managing public repositories from GitHub

Quick Start

An electronic version of this notebook is available at: http://hgwdev. soe.ucsc.edu/~miten/BookChapter_BioinformaticsBasics/Bioinformatics_basics.html.

Below is a brief summary of the steps you will perform during the exercise. If you need help with any of the steps, please see the instructions that follow in the exercise.

1. Download Anaconda (https://www.anaconda.com/download/)
2. Go to Terminal
3. Go to install software and follow directions to install marginAlign (https://github.com/benedictpaten/marginAlign)
4. Create virtual environment in marginAlign
5. Go to data set (http://hgwdev.soe.ucsc.edu/~miten/ BookChapter_BioinformaticsBasics/)
6. Copy link to downloadable files one at a time (files.zip and referenceEcoli.fasta.zip)
7. Download the zip file
8. Install poretools (https://poretools.readthedocs.io)
9. Install bwa (https://github.com/lh3/bwa)
10. Copy bwa to virtual environment executable path
11. Start the exercise.

Installing Anaconda

We will begin the exercise by installing Anaconda and some command line tools. For most Python applications, the Anaconda package is a good starting point. Most bioinformatics tools use Python 2.7, and Anaconda allows you to install Python tools and manage them yourself. (https://www.anaconda.com/download/).

Anaconda has some documentation to help troubleshoot. One of the goals of this exercise is to allow users to learn how to fix installation issues. The experience of troubleshooting will make it easier to install other bioinformatics software in the future.

Installing Software

Now that Anaconda is installed, we will install some specialized software for the MinION. We will begin with poretools (https://poretools.readthedocs.io/), which is a basic data extraction and QC utility. We will use BWA (https://github.com/lh3/bwa) as the alignment tool and then use marginAlign (https://github.com/benedictpaten/marginAlign) for doing some downstream alignment analysis. To install any of these tools, you can follow the instructions given on their webpages. Most of these tools also have their own set of instructions as well as interactive notebooks (like this one) that you can follow and use. Some instructions to help you get started are included in the workbook below.

Marginalign

Terminal is a standard and stand-alone tool on both MacOS and Linux-based systems. Within Terminal, marginAlign can be installed using the instructions provided on GitHub (https://github.com/benedictpaten/marginAlign). If you are just starting with Anaconda, you can use conda (Anaconda's package manager) to install different packages like git, pip, etc. You will see errors that will tell you which packages are missing. Once again, use Google as a resource when required.

Once you have installed marginAlign, also remember to create a virtualenv (instructions on marginAlign GitHub). Virtualenv is a virtual python wrapper that can install required packages within itself and then be used for the software. This is what marginAlign does using the virtual environment.

For the purposes of this exercise, we will work from the margin-Align folder. That is, we will install other tools and copy the data to use for the exercise in the marginAlign folder itself.

Test Dataset and Reference Files

Once the installations are complete, you will follow the instructions below and perform them step-by-step to see if you get the same results from the files provided on the links below (http://hgwdev.soe.ucsc. edu/~miten/BookChapter_BioinformaticsBasics/).

Downloading Test Data

When you go to the link above, you will find two zip files, files.zip and referenceEcoli.fasta.zip. Files.zip contains 140 fast5 files that are the output from a MinION run, and the referenceEcoli.fasta.zip contains the known reference sequence for a 4.5 megabase-sized bacterium named *Escherechia coli* (also referred to as *E. coli*).

You can click and download the zip files, and then unzip them once they are downloaded. Please note that we are doing this from the marginAlign folder.

Getting Started with UNIX Terminal

Generally, in a UNIX environment, like a MacOS or a Linux OS, you can install marginAlign, as well as other python utilities using Terminal. Terminal is a command line interpreter for bash, and you can read more about it online (https://en.wikipedia.org/wiki/Unix_shell).

What is Bash?

Bash is a UNIX shell where you can run Linux commands. You can read more at: https://en.wikipedia.org/wiki/Bash_(Unix_shell).

Running Commands

Once you have installed marginAlign, and also installed its virtual environment, you can run the commands in this notebook in two ways:

1. Continue using Terminal for the exercise, or
2. Use a Jupyter notebook to run the exercise in an interactive manner (preferable).

What is a Jupyter Notebook?

Jupyter notebook is an application that allows you to run command lines in an interactive manner. This lets you see the output of your commands real time and in the notebook itself. This also permits you to document your work, preserve notes for future usage, and share among collaborators. You can read more here: http://jupyter-notebook-beginner-guide.readthedocs.io/en/latest/what_is_jupyter.html.

You can copy the commands in this workbook, and then paste them in using a command line environment like Terminal.

NOTE

Please note that the symbol ! is used only to execute bash (see below) in the notebook. Alternatively, if you run these commands on command line, use them WITHOUT the !.

Using this Notebook

When you run these commands, you should expect to see the output shown in this link: http://hgwdev.soe.ucsc.edu/~miten/BookChapter_BioinformaticsBasics/.

Quick Notes for Jupyter Notebooks

1. When using a notebook, the way to execute a command in a cell (one of these blocks) is to either use a combination of shift + enter keys or click on the "Run Cells" button in the Cell tab above.

2. When you execute a command in a cell that takes longer to run, you will see an * on the left side of the cell in the square brackets instead of a number. This will go away once the command completes execution and you will see a number inside the square brackets, and also the output from that command below.

Let us Start the Exercise Now

What follows below in italics are comments that explain the commands we are going to execute. The commands themselves are not in italics and should copied and pasted into the notebook that you are working on.

 # Activate the marginAlign virtual environment (that contains dependencies)
 # (from marginAlign (https://github.com/benedictpaten/marginAlign))
!source env/bin/activate
 # This will activate the virtual environment within marginAlign. This means that
 # now the python being used is the one that you installed as part of the virtual
 # environment, along with marginAlign's python dependencies.

How to Install Poretools

(Remember that poretools is a basic data extraction and quality control (abbr. qc) utility.)

There are several ways to install poretools, including from GitHub (https://poretools.readthedocs.io/). An easy way to install poretools is to use pip and install it within the marginAlign virtual environment.

(Recall that pip is a type of Python package manager that allows you to install libraries that you will need for your work.)

 # Let's install poretools
!pip install poretools

 # Let's look at the help function
!poretools --help

Data Types

Most MinION data is provided in a sequence format called FASTQ which is written in all capital letters by convention (https://en.wikipedia.org/wiki/FASTQ_format). However, the MinION also provides the ionic current signal for each molecule when they are processed by the nanopore in a fast5 file. You can read more about fast5 files here (http://bioinformatics.cvr.ac.uk/blog/tag/fast5/).

QC tools like poretools can be used to extract sequence information from fast5 files. Alternatively, they can perform the same QC analyses on a FASTQ file if you already have one (more about this at the end of the tutorial).

Data Extraction and Basic QC

(Recall that the command lines below should be copied and pasted into your notebook.)

 # Extract FASTQ sequence from fast5 files
!poretools fastq files/ > reads.fastq

 # Let's make a read length histogram
!poretools hist files/ --saveas hist.png

 # Let's look at some basic statistics
!poretools stats files/

Sequence Alignment Using BWA

We use sequence alignment to measure similarity between the nanopore FASTQ sequences and a reference sequence. Burrow–Wheeler aligner (BWA) is a useful alignment tool for this purpose. Like the previous two software downloads, BWA is also available on GitHub (https://github.com/lh3/BWA). You can install it using the instructions on GitHub. Once installed, you can copy the BWA executable utility from the bwa folder into the marginAlign virtual environment.

```
# Lets copy bwa first from GitHub
!git clone https://github.com/lh3/bwa.git
```

```
# Go into the bwa folder
%cd bwa
```

```
# compile the software
!make
```

```
# Let's  copy the executable into the marginAlign virtual environment
# This will allow us to use it directly
%cp bwa ../env/bin/bwa
```

```
# now let's go to the main marginAlign directory
%cd ../
```

```
# Now let's run a sequence alignment using bwa
# Let's first look at help
!bwa
```

```
# We first need to index the reference genome
!bwa index referenceEcoli.fasta
```

```
# Now we can run the alignment using nanopore paremeters (-x ont2d)
!bwa mem -x ont2d referenceEcoli.fasta reads.fastq > alignment.sam
```

SAM and BAM Format

Sequence alignments are represented in a SAM (or BAM) file format: (https://samtools.github.io/hts-specs/SAMv1.pdf).

```
# Lets look at the first 4 lines of the SAM file
!head -n 4 alignment.sam
```

Alignment Statistics

Sequence alignment statistics can be used to measure the quality of read sequences as well as the alignment.

 # Now let's calculate some alignment statistics
 # using marginStats
 # Let's look up the help first
!bash marginStats -h

 # Now let's calculate the alignment identity
!bash marginStats alignment.sam reads.fastq referenceEcoli.fasta --alignmentIdentity

 # Note how the identity values are very low, this is because we are asking the question:
 # How identical are the reads to the reference that is 4.5 megabases in size?
 # Since our reads are not the size of the reference (look at the poretools stats output above),
 # what we should have asked instead to measure the quality of the read is:
 # How identical are the reads to the part of the reference that they align to?
 # In order to do this, we use a flag in marginStats called --localAlignment
!bash marginStats alignment.sam reads.fastq referenceEcoli.fasta --alignmentIdentity --localAlignment

 # Now that we have fixed the alignment issues, let's look at some more stats
!bash marginStats alignment.sam reads.fastq referenceEcoli.fasta --alignmentIdentity \
--readCoverage --mismatchesPerAlignedBase --deletionsPerReadBase \
--insertionsPerReadBase --readLength --localAlignment

 # Let's compute some stats and print values per read
!bash marginStats alignment.sam reads.fastq referenceEcoli.fasta --readLength \
--printValuePerReadAlignment --localAlignment

End of exercise!

Going Forward

Now you can have fun with data. Below are links to some datasets.

- Human genome DNA data (GM12878): https://github.com/nanopore-wgs-consortium/NA12878
- Human transcriptome RNA data (GM12878): https://github.com/nanopore-wgs-consortium/NA12878/blob/master/RNA.md
- *E. coli* (K12 MG1655): http://www.ebi.ac.uk/ena/data/view/PRJEB18053

Bioinformatics Resources

Here are some links to additional bioinformatics tools and resources:

MinION Bioinformatics Introduction

- PoreCamp: https://porecamp.github.io/2017/

Sequence Alignment

- minimap2: https://github.com/lh3/minimap2
- marginAlign: https://github.com/benedictpaten/marginAlign

De Novo Assembly

- Canu: https://github.com/marbl/canu

Alignment Statistics and Basic QC

- poretools: https://poretools.readthedocs.io/en/latest/
- poRe: https://github.com/mw55309/poRe_docs
- marginStats (part of marginAlign)

Chapter 10

Two Suggested Student Laboratory Exercises

Daniel Branton and Stephen Fleming

Two exercises described in this chapter are designed to be the introductory phases of a university laboratory course in which small groups of 2-4 students employ their own assigned instruments to grasp the foundations of nanopore sequencing. Because the best way to gain a full understanding of how nanopore proteins in a MinION or other nanopore sequencing apparatus self-insert and function in a lipid layer, the first exercise "Setting up a functional nanopore" puts into practice the information conveyed in chapters 1-4. Similarly, because determining an organism's sequence in a MinION requires the use of molecular engineering and bioinformatics tools, the second exercise "Identifying kimchi bacterial species by sequencing their DNA" puts into practice the information conveyed in chapters 5 through 9.

This chapter and the accompanying files that can be downloaded from https://www.worldscientific.com/worldscibooks/ 10.1142/10995#t=suppl describe and illustrate most of the specialized equipment and supplies needed for these two exercises. Machining and purchasing or borrowing the required used equipment can usually be affordably managed by a university teaching laboratory. While a laboratory course will often include a number of other associated topics and projects, the supply listings and explanations provided here are uniquely focused on the specialized equipment required for setting up a functioning nanopore, sensing single DNA molecules, and extracting, preparing, and sequencing an organism's DNA to identify its genome using a MinION nanopore sequencing instrument. Commonly found laboratory equipment such as spectrophotometers, dissecting microscopes, incubators, shakers, ring stands, and reagent grade chemicals, etc. are assumed to be available and are not listed here.

Setting up a Functional Nanopore

Because setting up a stable lipid bilayer and incorporating a single useable nanopore into this bilayer requires steady hands and some patient practice, it will usually take a novice at least 3 or 4 laboratory sessions to create a good lipid bilayer, insert a nanopore, and finally introduce analyte DNA, and record blockade events as single molecules are driven through the nanopore.

Terminology

DI water	Deionized or distilled water (preferably 18.2 MΩ cm).
PC	Diphytanoylphosphatidylcholine (1,2-diphytanoyl-*sn*-glycero-3-phosphocholine), the lipid that is used to make the bilayer membrane.
Patch stand	The Teflon piece in which a lipid bilayer that separates two volumes of buffer solution is formed and in which experiments assembling and using nanopores are conducted. (The word "patch" comes from research in which a patch of a cell membrane containing a channel adheres to a glass microelectrode. With proper amplification, the ionic currents through the channel can be monitored.)
Ag/AgCl	Silver/silver chloride, the material of choice for electrodes
"Cis" side	One of the two solution wells in the patch stand: the grounded side of the membrane, closest to the experimenter, where analyte molecules and nanopores are introduced and manipulated.

Teflon "plugs"	Small Teflon fittings, one of which holds and connects the *trans* side electrode to the head stage and three of which plug into the three access ports on the *cis* side, two of which are usually solution flow lines, and one of which holds the electrode that connects to ground.

General Setup and Operation

Measurements of ionic current through a nanopore in a lipid bilayer are obtained by applying a voltage across the lipid bilayer in an aqueous solution of 1 M potassium chloride (KCl) using silver–silver chloride (Ag/AgCl) electrodes. The suggested setup comprising a small homemade Faraday Cage, patch clamp current amplifier, digital-to-analog converter and a laptop computer occupies about 40 inches of laboratory bench space (Figure 10.1). The Faraday Cage, with a hinged top, encloses the Teflon patch stand in an aluminum holder, the current amplifier's head-stage, and the tubing and syringes that are used to smoothly exchange the *cis* chamber buffer with fresh buffer when required (Figure 10.1).

A separate Ag/AgCl electrode is in contact with each of the two volumes of buffer in the teflon patch stand. (See Figure 10.5 for all electrical connections). The buffer chamber closest to the operator is electrically grounded, and is termed the "cis side." The buffer chamber contacted by the electrode connected to the amplifier headstage of the current amplifier is termed the "trans side" (Figure 10.1 inset). Except during initial cleaning and buffer filling operations, all manipulations by the experimenter take place on the *cis* side.

The *cis* side and the *trans* side of the teflon patch stand are connected by a small, U-shaped tube, also made of Teflon (Figure 10.2). At the *trans* side, the tube is wide open. The opening of the tube in the *cis* side is terminated with a cone-shaped void in a Teflon piece with a very small (≤50 micron) aperture. This small aperture forms the only opening for an electrical connection between the *cis* and the *trans* side. Spreading a lipid bilayer across this aperture effectively blocks the flow of ionic current from one electrode to the other.

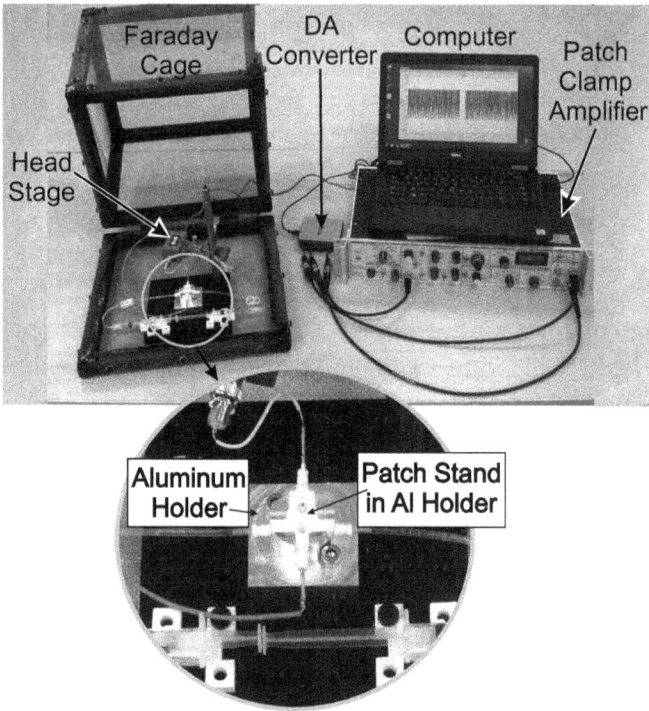

Figure 10.1. Components of a functioning nanopore apparatus. When closed, the grounded Faraday cage encloses the patch stand, head stage, and syringes for exchanging *cis* chamber solution. The circular Inset shows the Teflon patch stand in its Al holder screwed to the optical bread board into which is screwed a post to hold the head stage and on which the two syringes connected to the inlet and outlet of the patch stand *cis* chamber are also mounted. Note that the two syringes are mounted with their plungers in line such that if the two syringes have similar sizes, any withdrawal of solution from the *cis* chamber by pulling out the plunger from the left syringe adds the same volume of solution from the syringe on the right side. The Faraday cage is a home-built fully copper or aluminum screened wooden frame with a hinged top. The optical bread-board sits on the bottom screen of the Faraday cage.

Patch Stand Cleaning and Preparation

The patch stand must be thoroughly cleaned before and after each use to avoid problems with contamination.

It is best to have three syringes for this process, each outfitted with tubing that fits snugly over the end of the U-tube in the *trans* side.

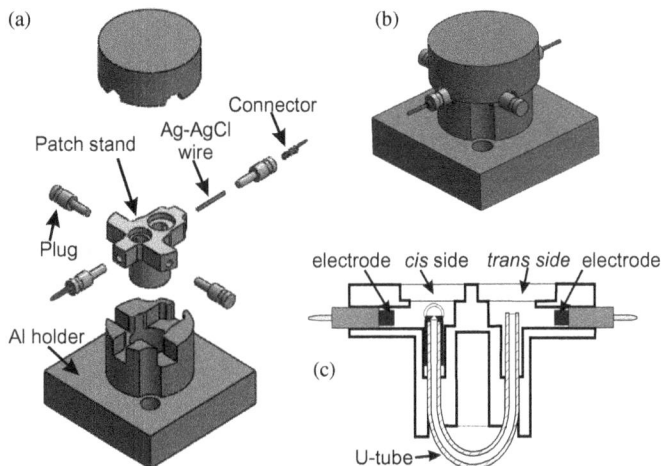

Figure 10.2. Teflon patch stand and its Al holder. (a) Unassembled components. (b) Assembled patch stand in its Al holder with inlet and outlet port plugged closed. (c) Patch stand cross-section showing Teflon U-tube connecting *cis* and *trans* compartments.

Always Clean the Setup Before use (Electrodes are not Involved in this Process)

The patch stand is normally stored in deionized (DI) water. Remove the patch stand from the water and shake off excess water. Insert the patch stand in its Al holder.

- Water

  Insert the three Teflon plugs into the three ports on the *cis* side. Fill the *cis* side with DI water. Attach the water syringe to the U-tube in the *trans* side and gently suck the water through the U-tube from the *cis* side until all the water is gone. Refill the *cis* side with DI water and repeat.

- Dry

  Remove the buffer syringe and the Teflon plugs and dry the patch stand with dry nitrogen. Attach the air syringe to the U-tube in the *trans* side and pull air through the U-tube until the U-tube is dry inside.

- Ethanol

  Dip the entire patch stand into a beaker of 200 proof ethanol and swirl. Remove patch stand from the ethanol and shake off excess ethanol. Insert the three Teflon plugs into the *cis* side. Fill the *cis* side with 200 proof ethanol. Attach the solvent syringe to the U-tube in the *trans* side and gently suck the ethanol through the U-tube from the *cis* side until all the ethanol is gone. Refill the *cis* side with ethanol and repeat.

- Dry

  Same as above.

- Hexane

  Fill the *cis* side with clean hexane. Attach the solvent syringe to the U-tube in the *trans* side and gently suck the hexane through the U-tube from the *cis* side until all the hexane is gone. Refill the *cis* side with hexane and repeat.

- Dry

Patch Stand Preparation: Lipid "Pre-paint"

Unless the clean Teflon aperture is first exposed to lipid, the membranes created across this aperture will be fragile and rupture prematurely. To enhance membrane stability, a lipid pre-paint step is required.

Start with a patch stand that has been cleaned according to the cleaning and preparation protocol above. The patch stand should be completely dry.

Lipid in hexane should be stored in the refrigerator.

Prepare Lipid in Hexane and Initial Patch Stand Setup

Add 200 μL hexane to a 0.4 mg tube of desiccated diphytanoyl phosphatidylcholine, which will be referred to as PC lipid, below. Gently mix by flicking the tube. It is helpful to dissolve newly ordered diphytanoyl phosphatidylcholine into hexane, apportion it out into multiple 0.4 mg samples per glass tube, which, after evaporating away the hexane with dry nitrogen, can be stored in small glass vials at $\leq$–20°C.

You will only use a small amount for each experiment, so the remainder in the glass vial can be stored in the refrigerator in a "humidor" of hexane, i.e. the small glass vial of PC/hexane should be stored in a larger glass vial containing ~1 cm of hexane and a tight-fitting lid.

Set up the patch stand under a dissecting microscope with a long working distance and look at the small aperture at the *cis* side end of the U-tube. Using a long pipette tip (gel-loading pipette tip), withdraw 5 μL of PC/hexane. Drop it onto the Teflon aperture, one drop at a time, allowing the solution to wick into the U-tube. As each drop is applied, give several seconds for a bit of the hexane to evaporate. After applying all 5 μL of PC/hexane to the aperture, attach the air syringe to the *trans* side of the U-tube and push air through the tube, from *trans* to *cis*, expelling any excess PC/hexane from the U-tube and drying it. Finally, allow at least 10 min for the patch stand to fully dry.

Preparation of the Ag/AgCl Electrodes

This should be done before each experiment.

Take a very small amount of bleach (just enough to cover exposed Ag wire) but not much of the Teflon — about 70 μL — in a 1.5 mL Eppendorf tube. Put the electrode into the bleach so that only the Ag/AgCl and a bit of insulating Teflon is submerged (Figure 10.3). Do this for both electrodes. Wait for 5 minutes while surface Ag is converted to AgCl. The color of fresh Ag should change from shiny silver to the matte dark brownish color of AgCl. If the matte dark brownish color is not uniform, sand down the surface of the electrode gently with fine grit sand paper, including the blunt tip of the wire. Clean and blow off any sanding dust, and retry.

Setting up the Patch Stand to Start an Experiment

After the patch stand has been pre-painted with PC lipid, set up the patch stand in its aluminum holder, with the Ag/AgCl electrodes and

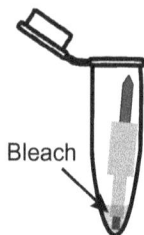

Figure 10.3.   Making Ag/AgCl electrode.

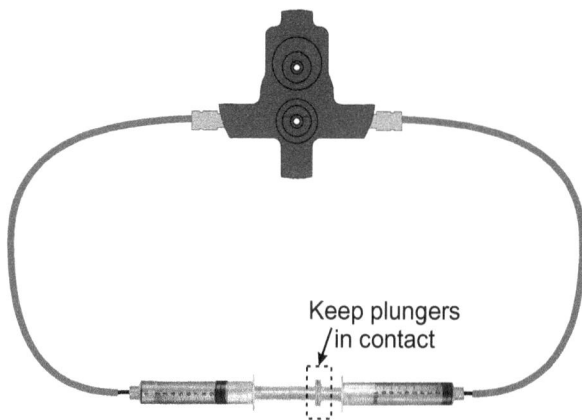

Figure 10.4.   Cis chamber perfusion connections.

the buffer-flow tubing connected to the patch stand as diagrammed in Figure 10.4. Choose a buffer to use. Typically, 1.00 M KCl, 10.0 mM HEPES buffer, at pH 8.00 is suggested. Pull some buffer into your buffer syringe (this is a clean syringe, not one used for the cleaning procedure), and use it to fill the U-tube from *trans* side to cis. Once the U-tube is filled with buffer and buffer has started to come through the *cis* side aperture, stop and remove the buffer syringe. If you will not be introducing nanopores, just put Teflon plugs into the two open patch stand ports on the *cis* side. If you will be introducing nanopores, fill the *cis* buffer flow lines with buffer using the syringes attached to each (one inlet, one outlet). Fill the *cis* and *trans* sides of the patch stand with buffer, ensuring there are no bubbles and that there is good contact between the solution and the electrodes.

At this point, the current amplifier will be overloaded. If even a small voltage is applied by the amplifier, the current should certainly overload. This indicates that ionic current is flowing freely, and the patch stand has been set up correctly. If you see zero current, there is likely to be a bubble present somewhere, either over one electrode, or on/under the aperture, or deep in the U-tube. Remove the bubble to achieve ionic current flow. A bubble in the U-tube can be removed by pushing fresh buffer *trans* to *cis* using the buffer syringe. To avoid air bubbles, it is helpful to briefly place the buffer under vacuum just before use with a water aspirator.

Creation of a Lipid Membrane

Remove a 0.4 mg/mL tube of PC lipid in hexadecane from −20°C freezer storage and allow the solid hexadecane to thaw at room temperature before use.

With a 10 μL pipettor and a long (gel-loading) pipette tip, dip the pipette tip into the PC/hexadecane. Do not suck up anything. A tiny bit of solution will wick into the pipette tip. Blow any solution out of the pipette tip (even if you do not see any visible solution in the pipette tip) into a Kim Wipe. Wipe off the outside of the pipette tip. There should now be nothing visible in or on the pipette tip. Nevertheless, a small amount of PC/hexadecane is still clinging to the inside of the pipette tip (even though you tried to get it all out) and that is all that you will need.

Dip this pipette tip into the *cis* side buffer (Figure 10.5(a)). Looking into the microscope while holding the tip under water, gently blow an air bubble out of the pipette tip (holding the tip under water), (Figure 10.5(b)). While it may seem that you have created a simple air-bubble, some of the PC/hexadecane from the inside of the pipette tip very probably has been extruded with the air and formed a monolayer around the bubble of air. While the bubble is still stuck to the pipette tip, rub the bubble across the aperture (Figure 10.5(c)). Watch the ionic current after each time that you rub the bubble across the aperture. Doing this a few times should establish a membrane bilayer whose very great electrical resistance prevents any detectable ionic current.

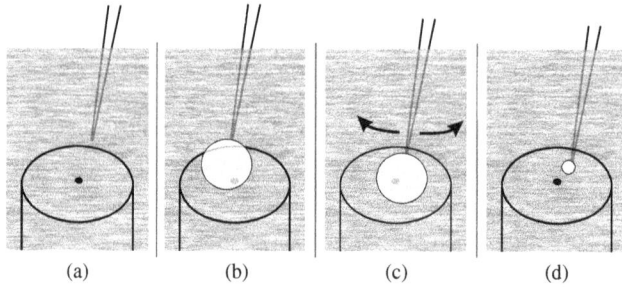

Figure 10.5.   Making a bilayer. (a) Dip a pipette into the *cis* side buffer. (b) Extrude air to create an air bubble. (c) Rub the bubble across the small aperture at the *cis* end of the patch stand's U-tube. (d) After forming a bilayer across the aperture, suck the air from the bubble back into the pipetter tip.

Once you establish a membrane that blocks ionic current, suck the air from the bubble back into the pipetter tip and remove the PC/hexadecane pipette tip from the *cis* chamber. Switch to a clean pipette tip for further work. If you have not established a membrane, or your membrane ruptures, switch back to the PC/hexadecane pipette tip, or prepare another PC/hexadecane pipette tip as described above.

Note: Too much of the hexadecane oil can entirely clog the aperture. Once the experiment is in this state, it is hard to remove the hexadecane, and the experiment must be cleaned up to start over. This is why such a minimal amount of hexadecane is strongly recommended!

Practice this step: Break the membrane and reform it. Use the "seal test" feature of the Data Acquisition software to look at the capacitive current spike that results when a brief voltage step is applied (Figure 10.6). A bilayer has a large capacitance. Keep forming membranes and try to get a high capacitance membrane with large current spikes (Figure 10.6(b)) that is likely to be a lipid bilayer extending across the full area of the aperture at the *cis* end of the patch stand's U-tube. Small-capacitance membranes (Figure 10.6(c)) are likely to be membranes that have formed with an excess of hexadecane and no true bilayer or only a very small area of bilayer into which it will be difficult to insert a nanopore. More bilayer area means higher capacitance, so break and reform new membranes and look for the size of the spike to increase. You can adjust the appearance of the spike using

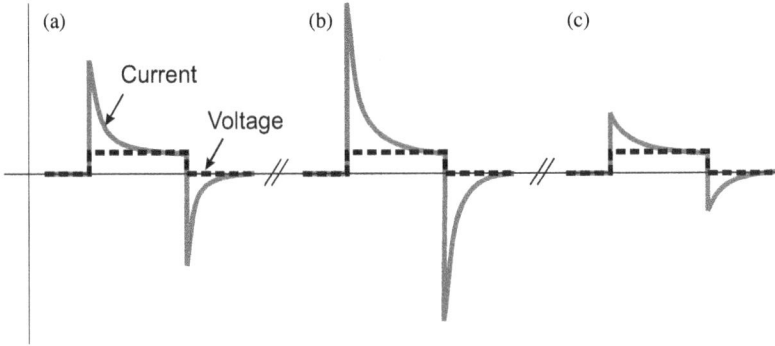

Figure 10.6. Capacitance spike examples. (a) A small area bilayer. (b) A larger area bilayer. (c) An excess of hexadecane; probably no bilayer area.

the "fast capacitance compensation" knob on the current amplifier as a way of keeping track of your highest-capacitance membrane.

Too much hexadecane oil can totally block the current by forming a useless plug of lipid across the aperture. This plug will not have a high capacitance and will make it impossible to insert a functional nanopore. You may be able to clear the oil out by pushing buffer through the U-tube (trans to cis) using the buffer syringe. If not, you will probably have to start over after having gone through the cleaning steps.

Inserting a Nanopore into a Lipid Bilayer

One you have formed a stable membrane, and you are reasonably sure it is a bilayer, it is time to try to insert a nanopore. Make sure you have the buffer perfusion lines connected, and buffer syringes set up and ready.

Take an aliquot of the cold alpha-hemolysin stock solution (2 microliters of the 10 ng/uL solution), and dilute it into 98 µL of buffer. Mix well, but be careful not to form bubbles.

Apply a voltage (typically + 120 mV) across the membrane. Take 2 µL of the diluted α-hemolysin, mix it into 20 µL of buffer, and carefully mix this solution into the *cis* chamber. Remove excess solution so that the *cis* side buffer is not bulging out of its well. Watch the measured ionic current, and wait for a nanopore to insert. One pore usually conducts ~120 pA when the voltage is set at 120 mV. When a pore inserts, immediately begin to gently perfuse buffer through the buffer perfusion syringes in order to flush excess hemolysin out of the *cis* chamber to prevent insertion of other pores.

After the *cis* chamber is thoroughly perfused, check to be certain the bilayer with its single nanopore is intact. The pore in an intact lipid bilayer should conduct a fairly steady current of ~120 pA when the voltage is set at 120 mV.

Troubleshooting

If after waiting ~15 min there is no sign of a nanopore having assembled and inserted itself through your membrane, i.e. there is no increase in the current, it is likely that your membrane is not a bilayer. It would be reasonable to flush out all pores from the *cis* side and try to reform a new bilayer. Pay attention to the capacitance of the membrane, and try to maximize it. When you are confident that your membrane is likely to be a bilayer, try to insert an α-hemolysin pore again.

If multiple pores insert very quickly, decrease the concentration of the pore solution you inject over the membrane after thoroughly perfusing buffer through the *cis* side and reforming the membrane. Unless you do a complete wash with acid, it is quite possible that you will leave a sufficient residue of α-hemolysin monomers in the setup so that even before you add any fresh α-hemolysin, you may see a pore inserting into the membrane you have just reformed.

Introducing Analyte Molecules

Prepare a solution of ssDNA of interest at a concentration of approximately 1 mM. This concentration can be varied. Use a pipette to inject about 10 μL of solution into the *cis* chamber and mix well, but be very gentle and careful not to disturb the fragile lipid bilayer with its nanopore. (A total of 10 picomoles of DNA in the *cis* side should provide a very high rate of molecule capture in the nanopore at voltages of 120 mV and above.)

Quite soon after introducing analyte molecules, they should start to transport through the nanopore as they are electrophoretically driven by the voltage bias that is applied across the lipid bilayer, usually 120 mV, *trans* side positive. As each molecule of DNA is traversing through the nanopore, the current through the pore will drop from ~120 pA to ~10 pA as shown in Figure 1.3(b). These changes through the nanopore — which indicate changes of the nanopore's conductivity — can be recorded on a computer using data acquisition software. A simple data acquisition program called "DataAcquisition" is provided and can be downloaded from https://www.worldscientific.com/worldscibooks/10.1142/10995#t=suppl.

Using the "DataAcquisition" Software

The data acquisition software, called DataAcquisition, is a Matlab program. To run the program, double-click on the Matlab icon on the computer desktop. Make sure the USB cable connecting the digital-to-analog converter (small gray-blue box) to the laptop is plugged in.

After the Matlab program has opened, navigate to the folder containing the DataAcquisition program. Type the following into the Matlab command line:

DataAcquisition

Then hit "Enter": This will start the program, and the rest is self-explanatory.

Cleaning After Experiment

Remove the electrodes and carefully rinse them in DI water. Spray them with ethanol, then rinse in distilled water and store them dry.

After an experiment, the patch stand and U-tube will be full of buffer. Rinse the patch stand thoroughly with DI water. Add DI water to *cis* side and pull this water through U-tube with the water syringe hooked to *trans* side until all water is gone, i.e. dry. Add hexane to *cis* side and mix with pipette to remove any remaining lipid and oil. Attach solvent syringe to *trans* side U-tube and pull hexane through U-tube until gone. Rinse patch stand with DI water. Add DI water to *cis* side and pull through U-tube using water syringe until U-tube is full of water. Stop. Remove water syringe, leaving the U-tube full of water.

Leave patch stand in DI water in a beaker covered with Parafilm.

Clean the buffer perfusion lines which may contain some residue of oil or α-hemolysin by flowing through DI water, then 10% bleach, then ethanol (190 proof is fine), then water, and allowing them to dry.

Intense Cleaning Procedure to Remove all Contamination

Place patch stand into a 200 mL beaker. Carefully add 50 mL of 10% nitric acid (diluted from concentrated nitric acid into DI water), which will just cover the patch stand. Leave overnight, or even better, after all the wells and tubing have filled with the nitric acid, bring the 10% nitric acid to a boil. Let it boil for 1–2 min, then after it has cooled rinse away all acid with distilled water, carefully dispensing the nitric acid (but not the patch stand!) into a large baking soda neutralization bath. Refill beaker with DI water and dispense into baking soda bath again. Repeat four times. Store patch stand and plugs covered with DI water in the beaker, and cover with Parafilm.

Supplies and Reagents

- Nitric acid, 10%
- Ethanol (200 proof)
- Hexane
- Nitrogen or compressed air (for drying)
- Commercial laundry bleach (usually 8% w/v NaOCl)
- 1 M KCl, 25 mM KH_2PO_4, pH 8.00
- ssDNA, polynucleotide of choice — should be ≥50 nucleotides
- 19 GaugePure (99.9%) Silver Wire (OD = 0.91 mm or 0.036 inches) (Rio Grande Jewelery)
- PTFE (Teflon) tubing, 1.4 mm OD (Component Supply Co. SWTT-21)
- PTFE\FEP heat shrinkable tubing, ~1.5 mm ID (Component Supply Co. SMDT-060)

- 1,2-Diphytanoyl-sn-glycero-3-phosphocholine — Also known as "diphytanoyl PC".
  Dissolve whatever amount of newly ordered diphytanoyl phosphatidylcholine into a convenient but known amount of hexane, and then apportion out all of this diphytanoyl PC in hexane into multiple samples each containing 0.4 mg of the diphytanoyl PC per glass tube. After evaporating away the hexane with dry nitrogen, store in small glass vials at $-20°C$ or colder. One of these glass tubes containing 0.4 mg diphytanoyl PC will be used by each group of 2–4 students after being dissolved in 1mL of hexadecane (N.B. not hexane) and stored at -20 in 1mL of hexadecane (N.B. not hexane) and stored at $-20°C$.
- Alpha-hemolysin
  Resuspend whatever amount of newly ordered freeze-dried powder of α-hemolysin to produce 1mg protein/mL in DI water. Add glycerol to bring to 50% by volume glycerol (creates 0.5 mg α-hemolysin/ml in 50% v/v glycerol). Aliquot 2 uL into each of multiple tubes. Snap freeze tubes in liquid nitrogen, and store at $-20°C$. For use in an experiment, dilute this 2 uL aliquot (at 0.5 mg/mL in 50% glycerol) by adding 98 ul of 1M KCl, and 10 mM HEPES, pH 8.00. This brings the α-hemolysin concentration to about 10 ng/uL. A few microliters (2–4) of this solution will be used by each group of 2–4 students to introduce nanopores into the lipid bilayer they have formed in the *cis* chamber.

Equipment

To minimize costs for a university, the patch clamp amplifiers and associated head stages listed below can often be borrowed or scavenged from equipment no longer used by laboratories that have moved on to next-generation equipment. Other than the digital–analog converters that are reasonably inexpensive, other items can be fabricated in a university carpentry shop and machine shop

1. Patch clamp amplifier and associated head stages
2. Digital–analog converters, e.g. National Instruments USB-6003
3. Laptop or desktop computers

4. Carpenter-built Faraday cages with Cu or Al screening
5. Machine shop-fabricated Teflon patch stand, aluminum-holder, and plastic syringe holders.

To facilitate fabricating multiple copies of components in a machine shop, files that contain machine shop drawings and SolidWorks CAD software files for use on computer-aided milling machines can be downloaded from https://www.worldscientific.com/worldscibooks/10.1142/10995#t=suppl. The folder labeled "PatchStand&HolderFigs" contains the following files: TeflonPatchStand, Plug, Electrode-Feedthru, AlPatchStandHolder, and OptionalPatchStandCover (needed only for lengthy experiments to prevent evaporation from the *cis* chamber). The files with PDF extensions show these parts as machine shop drawings that can be opened on any computer, whereas the corresponding SolidWorks files, with the SLDPRT file extension, can be opened on computers furnished with the Dassault Systèmes Corp SOLIDWORKS application and are for use on computer-aided milling machines. For example, the teflon patch stand drawn in the TeflonPatchStand.PDF file can be made in multiple copies using the machining instructions in the file TeflonPatchStand.SLDPRT.

The U-tubes for the patch stands (see Figure 10.2(c)) are best made and assembled into the patch stands ahead of student class meetings by the course instructor or teaching assistant. Using the instructions in Figures 10.1 and 10.2, producing each U-tube will require a 300–400 mm length of PTFE (Teflon) tubing, 1.4 mm OD (Component Supply Co. SWTT-21), and a 3–5 mm length of dual-shrink PTFE/FEP, ~1.5 mm ID (Component Supply Co. SMDT-060). A machined, sharply pointed ~150–200 mm long stainless steel rod, 0.8 mm diameter, (McMaster-Carr #8908K32) will serve as the mandrel to form a 20–40 μm diameter aperture through the PTFE/FEP that will cap the *cis* end of each U-tube. The most difficult aspect of the suggested procedure is finding a good machinist who can machine a clean sharp point on the mandrel. Producing a sharp point requires very careful machining, and even so the machinist should be asked to produce a sharp point on several mandrels so that the two or three mandrels with the best sharp points can be selected under a microscope. Also required is a rod or block of copper with one perfectly flat polished surface against which heat-treated malleable PTFE/FEP will be pushed.

To produce a U-tube, slide the mandrel into a 25-cm length of the Teflon tubing, leaving a short length of the mandrel's pointed end protruding from the tubing as show in Figure 10.7(a). Clamp the Teflon tubing and mandrel firmly in a horizontal position and then slide the PTFE/FEP heat-shrinkable tubing onto the Teflon tubing end as is also shown in Figure 10.7(a). It is best to clamp the shaft of the long Teflon tubing with the mandrel in it to the top of a linear ball slide (such as W.M. Berg Co LBSA-20) so that the end with the short length of heat shrinkable tubing can be smoothly moved to the center of a tungsten coil heater as, in Figures 10.7b) and 10.8(a).

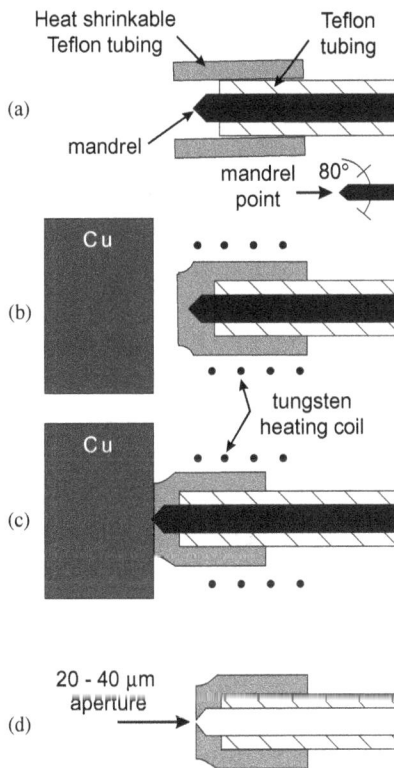

Figure 10.7.   Making a Patch Tube. See text for explanation of steps (a)–(d).

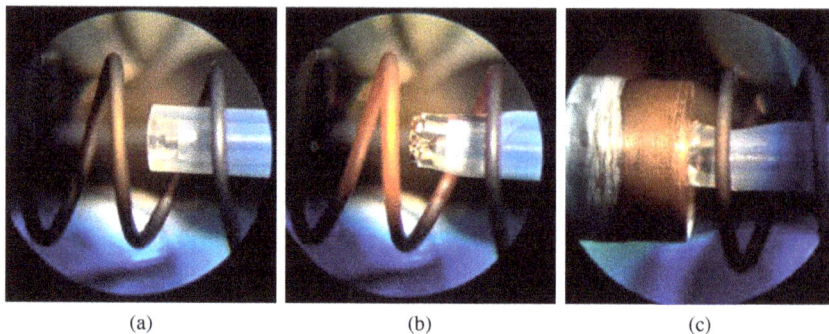

(a)   (b)   (c)

Figure 10.8.   Forming the 20–40 μm diameter aperture at the *cis* end of the patch stand U-tube. (a) After the heat shrinkable PTFE/FEP is slid into place (Figure 10.7(a)), this assembly at what will become the *cis* end of the patch stand U-tube, is moved to the center of a tungsten heating coil. (b) A voltage is briefly applied to heat the tungsten coil, shrink the PTFE/FEP and render it malleable. (c) The malleable PTFE/FEP is pierced by the pointed end of the mandrel as it is pushed and reformed against a smooth copper surface.

After the PTFE/FEP tubing is in place at the center of the tungsten coil, apply voltage to heat the tungsten coil and shrink the PTFE/FEP so that it slowly flows to make uniform contact around the mandrel point (Figures 10.7(b) and 10.8(b)). Then, as the voltage to the tungsten coil is turned off, quickly but gently push the linear ball slide to move the malleable PTFE/FEP and mandrel against the polished face of a Cu surface that is firmly held in place just beyond the end of the hot tungsten coil (Figures 10.7(c) and 10.8(c)). If the PTFE/FEP and mandrel are pushed against the copper with gentle single-finger force, the hard point of the mandrel will penetrate through the PTFE/FEP and slightly pierce into the softer than stainless steel Cu surface. This will produce a few microns deep indentation in the copper surface and a <50 μm diameter aperture through the PTFE/FEP. This aperture will be invisible to a human eye when the PTFE/FEP is withdrawn from the copper surface (Figure 10.7(d)). With a bit of practice, a 20–40 μm aperture, which should be verified in a microscope, can be consistently produced.

The flexibility of 1.4 mm Teflon tubing makes it difficult to push it through the 1.35 mm (0.053″) diameter channels leading from the

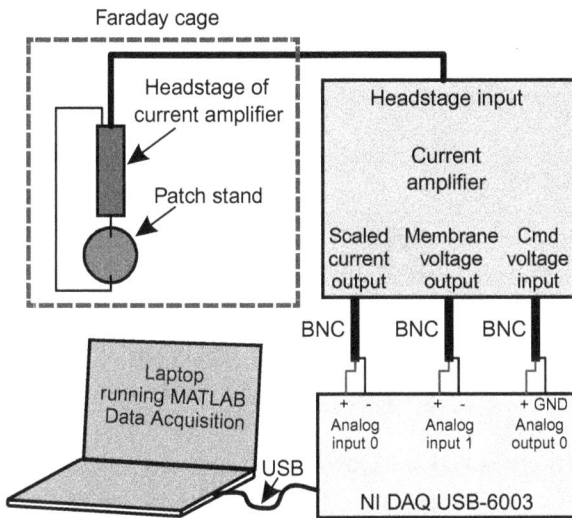

Figure 10.9. Wiring for connecting computer and DAQ to the patch-clamp amplifier and head stage.

bottom of the Teflon patch stand to the *cis* and *trans* wells. Instead, pull it through these channels after reducing the OD of the distal end of the Teflon tubing (the end without the heat shrinkable tubing) by stretching about 5–10 cm of its distal end after warming it with a hot air blower. The stretched portion of the tubing will have a reduced OD that can easily be slid into the 1.35 mm channels to serve as a leader strand so that the not-stretched portion can be pulled into and through the tightly hugging channels to form the U-tube.

The wiring connections from the patch stand through the headstage, patch-clamp amplifier, and DA converter to the USB3 input of the computer are shown in Figure 10.9.

Identifying Kimchi Bacterial Species by Sequencing Their DNA

Oxford Nanopore Technology's MinION is a robust, inexpensive sequencing instrument that now makes it possible for small groups of university students to identify an unknown microorganism by sequencing its DNA with their own assigned instrument. By identifying a bacterial species using a MinION, this exercise puts into practice the information conveyed in Chapters 5–9.

Supplies and Reagents

- Any of several varieties of fresh (not sterilized) kimchi available from a local Asian restaurant or food market.
- Gel electrophoresis supplies
- Phosphate-buffered saline (Lonza Cat#17-517Q)
- Lactobacilli MRS agar plates or MRS broth, agar and 10 cm petri dishes for making plates
- DNA extraction kits such as E.Z.N.A. Bacterial DNA Kit (Omega Bio-Tek Cat#D3350-01 and spin columns (or see Chapter 7 for alternate DNA isolation methods).

Equipment

- MinION and associated flow cells. Depending on anticipated ongoing MinION usage, ONT's Flongle adapters that permit usage of inexpensive small flow cells may be cost-effective.
- Tissue Homogenizer (*Ultra-Turrax, IKA Cat#40 268 00,* www.ika.net) and cartridges (*DT-20 Mischgefäß, IKA Cat#3703100*) or a small kitchen Blender.
- Centrifuge capable of >400 × g_n and volumes of 10–50 ml and appropriate centrifuge tubes.
- Centrifuge capable of ≥1000 × g_n and volumes of DNA spin columns

A good source of bacteria to identify is found in Kimchi, a type of fermented cabbaged-based food frequently consumed in Asia that some students may prepare themselves or that can be purchased fresh (not sterilized) for the course at a local Asian restaurant or food market. It is the result of spontaneous, but controlled, fermentation due to the development of a microbial community dominated by lactic acid bacteria present on the raw material entering in the preparation of kimchi (cabbage, radishes, leek, green onion, etc.). There are about six species of lactic acid bacteria commonly found in Kimchi including *Leuconostoc cutrium, Leuconostoc gelidum, Leuconostoc mesenteroides, lactobacillus sakei, Lactobacillus curvatus, and Weissella koreensis.*

Kimchi Processing

(1) Weigh out 5–10 g of kimchi. Be sure to include liquid component as well as solid, as this will facilitate homogenization.

(2) Using scissors or blade, mince solid pieces of kimchi into small (~2–3 mm) pieces.

(3) Homogenize kimchi with a tissue homogenizer or kitchen blender. Add phosphate-buffered saline (PBS) as needed to facilitate homogenization. The result should be a slurry with solid pieces <1 mm.

(4) Transfer homogenate into centrifuge tube and centrifuge at 400 g for 2 min.

(5) Visually adjust supernatants to the same volumes as the sedimented material. Remove the supernatants and save. Discard the sedimented material.

(6) Dilute supernatant 1:100 in PBS.

Because the diluted supernatant will almost certainly contain several different species of lactic acid bacteria which will complicate species identification for a student who has not previously identified a bacterium by sequencing its genome, it is best to plate out several further dilutions of the supernatant on a solid medium appropriate for the growth of the most common fermentative bacteria found in kimchi. The plating will lead to individual colonies each derived from a single bacterial species. Individual colonies can then be selected, grown in liquid medium so that the DNA of a single species can be extracted, purified, sequenced, and then matched to databases that are available for the lactic acid bacteria commonly found in kimchi.

Proceed as follows:

Step 1: Spread kimchi homogenate on MRS agar plates. Depending on how the kimchi was prepared, several dilutions on separate plates will be necessary (1:500, 1:1000, 1:10,000) to produce distinct individual colonies, and volume.

Step 2: Incubate overnight at 25°C.

Step 3: Pick several colonies, using a loop or sterile toothpicks.

Step 4: Transfer each picked colony to 3 mL MRS broth in culture tube.

Step 5: Incubate on shaker at ~200 rpm overnight at 25°C.

Step 6: Perform genomic DNA extraction using EZNA Bacterial DNA Kit or equivalent (see Chapter 7).

After the genomic DNA of a single species is purified, prepare the DNA for nanopore sequencing by ligating the leader strand, enzyme motor, and tethering element (see Chapter 8) following the instructions for the selected Oxford Nanopore Technologies sequencing adapter kit. Then determine the unknown species' DNA sequence in a MinION using the instructions provided with the MinION.

Optionally, the identification of the bacterial community found by matching its genomic sequence to databases that are available for the lactic acid bacteria can be confirmed by the instructor or the students themselves by amplifying the DNA using PCR with species-specific primer sequences found in the genes coding for the bacterial 16S ribosomal RNA. The amplified material can then be checked for the presence of the appropriate base-pair PCR product using electrophoresis on a 1% agarose gel. The sequence of the pairs of primers used in the different PCR reactions as well as the expected base-pair products are listed below.

Leuconostoc **Species**

Kim B.J., *The Journal of Microbiol.*, 2000, 132–136
L. gelidum "LGel" — 1290bp
F: 5′ TCGTATTCGTATCGCATGTCGTATCGCAT 3′
R: 5′ TAGACGGTTCCCTCCTTAC 3′
L. mesenteroides "LMes" — 1150bp
F: 5′ AACTTAGTGTCGGATGAC 3′
R: 5′ AGTCGAGTTACAGACTACAA 3′

L. *citreum* "LCit" — 1298bp
F: 5′ AAAACTTAGTATCGCATGATATC 3′
R: 5′ CTTAGACGACTCCCTCCCG 3′

Lactobacillus Species

J. Lee *et al.*, *Journal of Microbiological Methods* 59 (2004) 1–6
L. *curvatus* "LCur" — 400bp
F: 5′ GAGCTTGCTCCTCATTGATAA 3′
R: 5′ TTGGATACCGTCACTACCTG 3′
L. *sakei* "LSak" —
F: 5′ GATAAACCAAATGTGTAGGG 3′
R: 5′ TTGGATACCGTCACTACCTG 3′

Weissella Species

J. Jang *et al.*, *EMS Microbiology Letters* 212 (2002) 29–34
"Weis" — 725bp
F: 5′ CGTGGGAAACCTACCTCTTA 3′
R: 5′ CCCTCAAACATCTAGCAC 3′

Guide to On-line Files.

1. The movie showing how the Dda helicase moves to split dsDNA into ssDNA that can be downloaded from https://www.world scientific.com/worldscibooks/10.1142/10995#t=suppl runs best in the VLC Player program. VLC is a free download from the Web and is available for both PCs and Macs.

2. The files that contain the computer driven milling machine instructions that can be downloaded from https://www.world scientific.com/worldscibooks/10.1142/10995#t=suppl cannot be opened on most computers unless the computer contains the specialized drawing program for making 3-D drawings and the corresponding automatic mill instructions. The files are made available so that the user can give them to a machine shop that has computer operated milling machines.

Index

Figures and tables are indicated by *f* and *t* after the page number respectively.

www.ingramcontent.com/pod-product-compliance
Lightning Source LLC
Chambersburg PA
CBHW050601190326
41458CB00007B/2131

* 9 7 8 9 8 1 3 2 7 0 6 0 2 *